PRACTICAL ASPECTS OF DATA COMMUNICATIONS

PRACTICAL ASPECTS OF DATA COMMUNICATIONS

Paul S. Kreager
Computing Service Center
Washington State University
Pullman, Washington

McGRAW-HILL BOOK COMPANY

New York   St. Louis   San Francisco   Auckland
Bogatá   Hamburg   Johannesburg   London   Madrid
Mexico   Montreal   New Delhi   Panama   Paris
São Paulo   Singapore   Sydney   Tokyo   Toronto

Library of Congress Cataloging in Publication Data
Kreager, Paul S.
  Practical aspects of data communications.

  1. Data transmission systems. 2. Computer
networks. I. Title.
TK5105.K73   1983   621.38   82-12552
ISBN 0-07-035429-4

Copyright © 1983 by McGraw-Hill, Inc. All rights reserved. Printed in the United States of America. Except as permitted under the United States Copyright Act of 1976, no part of this publication may be reproduced or distributed in any form or by any means, or stored in a data base or retrieval system, without the prior written permission of the publisher.

1234567890   DOCDOC   89876543

ISBN 0-07-035429-4

The editors for this book were Stephen Guty, Virginia Fechtmann Blair, and Charles P. Ray; the designer was Mark E. Safran, and the production supervisor was Teresa F. Leaden. It was composed by Lorraine Spence and Joseph Friedman in the McGraw-Hill Book Company Publishing Technology Department and typeset in Times Roman by Benjamin H. Tyrrel, Inc.

Printed and bound by R. R. Donnelley & Sons Company.

To Etta and Vernie Rees, two great people.

CONTENTS

Preface — xi

CHAPTER 1 WHAT PRACTICAL ASPECTS?   1

Mini-Glossary — 2
Major Areas of Concern — 4
Scope and Goals — 7
Chapter Topics — 9

CHAPTER 2 LOCAL NETWORKING   13

Media — 14
Implementation Concepts — 18

CHAPTER 3 FACILITY DESIGN   27

Terminal-End Considerations — 29
Hallway Cabling Requirements — 34
Vertical Cable Access — 40
Interbuilding Considerations — 42
Communications Area Considerations — 45

CHAPTER 4 INTEGRATING THE NETWORK AND FACILITY   47

Feeder Cable Concepts and Practices — 48
Bulk Feeder Cabling Installation — 50
Terminal-End Topics — 57
Facility and Network Labeling — 62
Why All the Network and Facility Bother? — 68

CHAPTER 5 COMMUNICATIONS AREA IMPLEMENTATION TOPICS 71

Communications Area Facility Design 72
Communications Area Features 80

CHAPTER 6 SPECIAL DATA CIRCUIT IMPLEMENTATION TOPICS 89

No-Modem Local Data Circuit Implementation 90
Coaxial Local Data Circuit Implementation 95
Cable Assemblies 97
Adapter Assemblies 102
Miscellaneous Circuits 106

CHAPTER 7 CONVENTIONS AND PRACTICES 109

Color Coding 110
EIA Mnemonics 114
Punch Block Panel Wiring Conventions 114
Punch Block Naming and Labeling Conventions 115
Punch Block Tab Numbering Conventions 119
SEND/RECV Conventions 120
Floor Tile Cutting Standards 122
Miscellaneous Practices 123

CHAPTER 8 DOCUMENTATION 125

The Data Communications Reference Library 127
Operational Documentation 131
Network Documentation 136
Miscellaneous Documentation 149

CHAPTER 9 RELATED TOPICS 157

Support Personnel 158
Maintenance Program 162
Management Issues 167

APPENDIX 173

Networking Checklist 174
Facility Checklist 177

Documentation Checklist	180
Custom Projects—Construction Notes	180
Parts Identification and Vendor References	185
Vendor Information	194

INDEX 197

PREFACE

In my visits over the years to various data processing facilities I have always been impressed by the lack of attention given to the practical day-to-day details associated with data communications. In the facility itself a great deal of time and frustration is spent continually working around built-in problems: difficulties in running cable to terminals, forever changing the power distribution to support equipment, offices not designed for data, etc. There is seldom any coherent plan for networking within the facility, as is demonstrated by the wires still being pulled one at a time with little thought beyond the present need. The computer terminal wiring generally is difficult to trace, is not documented, and shows little sign that commonly accepted conventions and practices are in use. Interbuilding networking shows the same general problems.

In data circuit implementations, some problems are overcome at great expense and complexity because of a lack of knowledge at the technical level of networking: modems are in use where they are unnecessary; black boxes are in use instead of simple connector adapters; and so on. These problems at the facility and network implementation level have arisen during the normal evolution of the data processing business, and many installations do not have staff trained at the engineering level to handle them. Typically, decisions are made by people out of the software ranks who simply are not aware of the alternatives. Implementation also is done by software people who are not trained at the hardware level.

Today the data processing environment is changing dramatically. Distributed processing, local networking, and word processing are literally exploding in use. The old machine room has been partitioned by the growing importance and complexity of its attached network. This creates new and special problems specific to data communications. The remainder of the facility outside the central processing area is approaching one terminal per office in many applications. Terminal clustering also is common. Local networking involves a rich mixture of wire-pair and coaxial circuits, with fiber in some applications. Even buildings having no central computer complex now require computer terminals on a large scale.

In addition to facility design and its related network implementation, there are many details necessary to the successful operation of the data communication function. The data communication facility and systems must be dynamic to respond to everyday changes, new equipment, new networking schemes and architectures, and evolutionary changes in the media. This requires policies, procedures, meaningful documentation, trained personnel, and proper management direction. These topics, then, form the basis of this book: the practical side of data communications. To my knowledge, no other book is devoted to this particular information. The content is derived primarily from personal experi-experiences, and it certainly does not cover all the tricks of the trade. However, it is at least a starting point for centrally locating some of this knowledge and making it available to others. I would greatly appreciate feedback, suggestions, and personal experiences of others to add to any future editions of this book. This information should be useful to almost everyone involved in data commu-communications, from the network technician to the engineers and managers whose livelihood is in the data communications environment. It is my desire that readers find in this book practical solutions to data communication problems and alternative viewpoints and implementations to enhance and improve their own operations. The solutions presented here may not be elegant, but they are useful in the practical world most of us have to work in.

Before closing the preface, I would like to give thanks to those who supported this book and made it possible. The vast majority of the photos used in the text were taken at the Washington State University Computing Service Center, and the how-to-do-it flavor of the book would have been very difficult to accomplish without them. A few photos of other facilities of the University add to the value of the material. The manuscript was developed and generated by using the text preparation and printing facilities of the Computing Center. This greatly decreased the time necessary to prepare the manuscript in comparison with the old cut-and-paste method of writing. A large thanks is due WSUCSC.

It is still true that no person is an island; I too have profited from my association with colleagues over the years. In this regard, I would like to acknowledge John Sobolewski and Lynn Cannon for the valuable experiences we have shared. And lastly, many of my weaknesses are in the good old rules of grammar and related skills which are so necessary to writing a book. My thanks to my wife, Mary Jane, whose expertise and help in this area is much appreciated.

Paul S. Kreager

PRACTICAL ASPECTS OF DATA COMMUNICATIONS

WHAT PRACTICAL ASPECTS?

CHAPTER 1

2 WHAT PRACTICAL ASPECTS?

Current literature on data communications does a good job of keeping us abreast of the latest products, methods, and theory. This information helps us modernize our equipment base, decide among the various implementation possibilities, and understand how things work technically. It is certainly practical, because we do use it to go about our everyday chores in data communications. But practical data communications, in that sense, stops short of the real down-to-earth stuff like working with the mess under the pretty raised floors, how to go about wiring a building, and what kind of parts to use for building data cables.

That kind of practical information is rarely found in the literature; it is assumed to exist in your shop. Those of us in the business know, however, that it is very seldom available in our shops. It always seems to be learned the hard way, by experience. Staffing plays a large role in how data systems are implemented at a practical level. You can see the difference between shops which have hardware staff support and those which do not, i.e., between having people who know about the nitty-gritty details of practical data communications at the basic implementation level and not having those people. These, then, are the practical aspects of this book and the level of discussion in the following chapters.

MINI-GLOSSARY

Some of the jargon used in the text may be unfamiliar to some readers, so before going any further, let's look at the buzzwords. Readers who are already familiar with the terminology may only want to scan the mini-glossary to find how certain terms are used in this book. The definitions are not precise or detailed; they are kept at the normal working-level parlance.

327X Designation of a line of equipment manufactured by IBM and other plug-compatible vendors which includes terminals and controllers. The equipment itself is not of much interest here, but its communication media is. Coaxial cable is used to connect this equipment to the computer and, more specifically, RG 62A/U coaxial cable.

A-E Short for architect-engineer.

asynchronous (async) Also called start-stop communications. Crudely, async communication is done without bit clocking or timing information being transmitted with the data. Data is extracted from the information stream by knowing when a block of bits (character) begins and ends, usually by adding start and stop bits to the block. This is in contrast to synchronous communications, where timing information is imbedded with the data.

AWG American wire gauge. Usually suffixed to a number to show which wire standard is being used; for example, 22 AWG refers to number 22 wire for that standard.

bandwidth Formally, the range of frequencies a communication channel will pass. Loosely, how much information, i.e., bits per second, you can send through a channel.

channel   Here, a communications path for data.

coax   Short for coaxial and generally meaning coaxial cable, even though the word "cable" may be missing.

comm   Short for communication(s).

conduit   As used here, it is either a plastic or metal tube through which cable is passed.

controller   Here, a device attached to a computer through which terminals access the computer, i.e., a communication controller.

crosstalk   Signals (or noise) coupled from one communication channel into another. Audio crosstalk is occasionally heard over the telephone when you hear some other party's conversation. Here, crosstalk refers to one data cable radiating its signal to another, generally adjacent, cable.

data comm (datacomm)   Short for data communication(s).

distributed processing   Loosely, processing power which is being distributed geographically rather than being concentrated into one central computer site. Mini- and microcomputers, for example, can be attached remotely to a midi- or maxicomputer to form a distributed processing network. A smart terminal may off-load some of the processing usually done entirely by the central processor. That would be a form of distributed processing.

DP   Short for data processing.

EIA   Abbreviation for Electronic Industries Association. "EIA" is used loosely to identify the type of physical connection and electrical protocol used by certain devices. Usage is very loose; e.g., "EIA connector" refers to connectors which meet EIA specification RS-232.

FEP   Abbreviation for front-end processor.

front end   The communication controller connected to a host processor and through which various computer terminals and other user devices are attached.

gen   Short for "generation." Most often used in reference to a software generation, where a change is made to the operational software.

Halon   A trade name of Du Pont for a chemical used to extinguish fires. Halon is commonly used in computer facilities.

handshaking   A term used in reference to how devices "talk" to one another, e.g., how a terminal exchanges information with a computer and keeps itself in synchronization. Handshaking can refer to either or both the electrical or character-level protocol.

hardwired   The direct connection of devices with only wire, i.e., with no intervening devices. For example, a hardwired terminal would be one directly connected to a communication controller rather than through modems.

host   Loosely and generally, a large central computer (site) which offers various timesharing and other services.

link   A communications path, e.g., a telephone line connecting a terminal to a remote computer.

long-haul   In contrast to short-haul, data communications over an extended distance such as city to city. A leased telephone line connecting a terminal to a computer in another city would be a long-haul line.

machine room   Same as computer room.

modem   Acronym for modulator/demodulator. A device which modulates a data signal with a carrier before sending it over a communication path and, conversely, demodulates the data from the carrier when receiving information from the communication path.

PM   Abbreviation for preventive maintenance.

port   An entry point into a data communications or computer system. For instance, the communication controller linked to a computer has many ports, or sockets, into which terminals are attached.

POS   Abbreviation for point of sale, e.g., POS terminals.

protocol   A set of conventions or rules by which devices communicate, or "handshake." There are often several layers of protocol. Each layer defines a certain aspect of communications: physical, electrical, character level, or even session level.

PVC   Abbreviation for polyvinylchloride, a plastic. Of concern here is that PVC is used to insulate wiring.

raised floor   A floor system which is raised above the principal support flooring to form a cavity useful for installing power and signal cables for a computer system which rests on the floor surface.

short-haul   Generally, and in contrast to long-haul, data communication paths within a small region (city) and covering distances of only a few miles. A line connecting a terminal in one building to a computer across town would be a short-haul line.

stranded (wire)   A wire which is made up of many finer wires, i.e., strands.

synchronous (sync)   Data communication in which timing information is transmitted with the data. Generally, in contrast to asynchronous communications.

Telco   Short for telephone company.

terminal   Here, a device at the user end of the communication link, e.g., a CRT terminal.

TP   Abbreviation for teleprocessing.

UPS   Abbreviation for uninterruptible power system.

WP   Abbreviation for word processing.

MAJOR AREAS OF CONCERN

What is it that we will be concerned about in this book? There are many things: the facility, the network, and the smaller things too, which help or hinder the way we go about our business. For a moment now, let's mentally remove ourselves from our routine work environment, try to look back at it from an unbiased viewpoint and ask a few questions about each of the major areas of concern:

Local Networking

- When a new terminal is required in the building, how do we go about installing it? Do we think ahead to when the next terminal after this one will need installation, or do we simply add terminals one at a time as needed? Economies of both time and money can be realized by planning ahead a little.
- Does the physical environment make it difficult to string cable for a new terminal installation? Can a variety of media be handled? Could a little work invested in the facility now make things easier in the future?
- Where a building network exists, is it dependable? Is it predictable in performance? Is it documented? Is it a mess? Can you troubleshoot it?

Facility

- Does the facility have horizontal and vertical cable paths to allow efficient local networking?
- Are the offices designed for data? Can terminals be easily installed in the offices? Can they operate from either a dedicated or a dial line? Will the terminal, modem, telephone, and power needs adapt readily to people who are changing their office habits, like moving furniture?
- What happens when a different communication medium is required? Does switching from twisted pair to coax create problems? When service requirements expand, will changing a 25-pair cable to a 100-pair one be difficult?
- When new equipment is added, do you always seem to need a change in the power distribution? Costly, isn't it?
- Ever think of standardizing the cutting of raised-floor tiles? Sometimes it is better to fit the equipment to the tile rather than the other way around. Sure saves on the tile budget.

Components

- Is it difficult to add or change equipment?
- Does it become confusing with the variety of cable types and adapters necessary to implement data communication circuits?
- Are the equipment enclosures and racks practical? Or do they just look pretty?

Documentation

- Are customer and vendor agreements readily available to the network staff? Does the staff understand them?
- Do you have archive documentation available for statistical purposes?

- Are your equipment and network components labeled consistently and in a meaningful way?
- Can you trace a circuit from one end to the other on your documentation? How many times have you wished you could?
- When you have a really nasty network problem, do you have ready access to the technical information required to solve it?
- On long jobs, lasting several days to weeks, can you tell where you are in the job? What needs to be done to complete it? What are you waiting for? Think about this for installing a new circuit with its associated modems, terminals, host interface requirements, hardware/software generation (gen) requirements, etc.

Operations Area

- Are operating procedures clear to everyone? Can backup personnel understand them?
- Are Mondays terrible? Why does everything go to pot over the weekend and show up Monday morning?
- You have an intermittent network problem and run to the hard copy for messages or clues. You find all messages are on the printer roller because of a lack of paper. Do similar problems indicate unclear lines of responsibility?
- Can you pinpoint some problems to work shift changes? Is information being carried over from shift to shift properly?

Management

- Are lines of responsibility clearly drawn from the network supervisor on up? Can the administrative boss circumvent the network supervisor and cause problems through a lack of technical understanding?
- Do you always seem to be running into dead ends because of management and administrative delays, unresponsiveness, and misunderstandings?
- Networking says it must be a host gen problem. Systems says it can't be them! Is finger-pointing a problem?
- Are the agreements made by management clear as to their impact on working relationships between the customer and the network staff?
- Are the various support groups uncoordinated?
- Is the path clear for acquiring normal equipment additions to support a growing network?

If you don't have any of the preceding problems, you should have written this book years ago! It would have spared most of us a lot of frustration. On the other hand, many of these problems might be examples right out of your shop. Maybe something could be gained by spending some time addressing them. Most of the reasons for them can be put into the following categories:

- Poorly designed facility
- Poorly designed network
- Ineffective procedures and policies
- Lack of planning
- Lack of decisions
- Inadequate information or documentation
- Unclear lines of responsibility
- Management not taking its share of the responsibility

This book will not solve all those problems; there will always be problems. However, it will attempt to provide some solutions, show some alternatives, and offer practical advice based on job-related experience in addition to classroom learning. It is hoped that this information will keep the fires burning low and prevent major flare-ups.

SCOPE AND GOALS

Data communications is different things to different people. To some people it is satellite transmission; to others it is high-bandwidth CPU communications; and to still others it may involve a public data network. Some people concern themselves only with digital data communications, some only with analog, and others with both forms. One person may focus on serial data transmission, whereas parallel transmission is the focus of another person. In short, there is a tremendous variety of subjects within data communications, and it is important to define more precisely what areas are covered in this book.

The people who are addressed by this book work in typical computing center situations where a computer service is provided to a terminal user via a network connecting the two parties. In some cases, the environment may be a pure network house with the computer service physically quite remote but still accessible via the network. Of prime interest is the overall local environment of the people in this situation. This includes the immediate facility, the people within it, and the mechanisms in place to make it function as a data comm shop. The facility may house computers, front ends, modems, terminals, remote batch, Telco links to the outside, local links internally and to adjacent buildings, word processing equipment, distributed data processing equipment, or any of many similar devices. The networks within the facility are principally implemented with common twisted pair, coaxial cable, or, in many cases, both. Serial data transmission is the main mode of communication, and it is used with both synchronous and asynchronous terminals.

In scope, the information presented generally consists of data comm topics common to most DP-type organizations. At the facility level, the interest is held to areas of immediate concern to the network function, i.e., how to design and build facilities to overcome wiring and other problems related to the network. Other facility items, such as security and Halon fire protection, are not germane. Likewise, discussions of networking are held to areas which were described previously at a practical level and are of common interest. No space is devoted to other networking schemes, such as radio broadcast, infrared transmission, or more specialized forms of data communications. Technical detail is held to a minimum. The information should be readily understandable to almost anyone whose livelihood is in data communications.

The overriding goal of this book is to show how to do data comm at the everyday level. The intent is purely practical with little theory. It is recognized that keeping the subject to this level forces some readers to review methods already in place in their own shops. The more experienced people would tend to have already developed their own methods and procedures similar to those presented here. Even in these situations, alternative viewpoints and implementation schemes which will be of value to these people are likely to be presented here.

Many tricks of the trade which are included are acquired only through experience. It has been my observation that many of the people working in data comm have come from the software ranks. This can be a handicap, because data comm requires knowledge outside that acquired as a programmer. Both hardware and software people are needed to provide really effective and knowledgeable data comm support. Typically, however, the hardware side is missing. Because I am in the hardware ranks, I present information from the engineer's viewpoint. This, supplemented with the reader's software training, hopefully will provide a better balance of the knowledge necessary to work effectively in data communications. In several cases, the philosophy behind the way things are done is included to further solidify the reasoning. It's surprising how many people never take the time to ask themselves why they are doing something in a particular way. Many of them do it simply because they were told to—never questioning the reasons why. The philosophy is important to enhance a person's understanding of the subject.

To some, the subject matter of this book may seem out of date because of the heavy use of twisted pair and RS-232 topics. Please keep in mind that this book is not a review of new technologies and how to use and deal with them. We are interested here in proven ways to go about our daily tasks in data communications. That is why the material was chosen; it should be familiar to most people practicing our art. Extensive discussion of RS-422, RS-423, RS-449, DDS, and the like would have detracted from the points of real interest; readers would spend too much time worrying about their unfamiliarity with, say, RS-449. Good ideas stand up against time. I think the philosophy behind most of the topics here applies to state-of-the-art data communications, as well as to older techniques.

CHAPTER TOPICS

Chapter 2 discusses local networking in general and then proceeds at a conceptual level into areas of particular interest to people in data processing facilities. The subject of local networking is very broad. Each person you ask for a definition of a local network (local area network, or LAN) is likely to give you a different answer. Very basically, "local networks" as used here may be considered to mean inter- and intrabuilding networks up to a campus level. We are interested in how networks are implemented within facilities, or between adjacent facilities, to connect common terminals to data processing systems. Installation is via common twisted pair and coaxial cable along with Telco short-haul circuits in some applications. Data transmission is primarily serial. We are not considering Ethernets, CPU channels, or anything else more advanced than the basic twisted pair and coax circuits. The goal of this chapter is to show, in general, what local nets are and conceptually how they are implemented into data processing facilities.

Chapter 3 switches somewhat abruptly to the subject of the data processing facility and how such a facility should be designed. This area is terribly neglected, as is amply demonstrated by most facilities. Very few facilities are properly designed for data. Those of us who work in a building designed specifically for computer and terminal usage are fortunate indeed. Most computing centers have rooms designed to house their computer mainframe and peripheral equipment but stop short of providing wiring facilities for the entire building. The situation is probably worse yet for most buildings having large numbers of terminals and no computer center.

The difficulties of installing a local network under these circumstances generally explain why many networks are implemented with little planning for the overall needs. It is frustrating enough to get just one terminal wired, let alone worrying about future terminals. Cost also can become a significant factor. Rather than invest considerable money to make the building wireable, circuits are wired one at a time as needed. Unfortunately, this approach generally is more costly in the long run, continues the frustration, and makes it very difficult to manage, change, and maintain the resulting local network.

The main point of this chapter is to suggest that it is well worth the effort to put a little thought into how the building should be utilized or modified to support a local network. This is true for old as well as new structures, and it becomes more important as the number of terminals grows. Knowledge of Chapter 3 will lead to flexible facilities for the data processing environment. People doing the interface work with architects-engineers for their own facility should find this information especially useful.

Some detail about networks and facilities having been provided by Chapters 2 and 3, Chapter 4 proceeds to tie the two subjects together. Detailed information is presented to show how the network is installed into the facility. The result is a facility designed and implemented for data. In the past, these two areas were always considered separately. Most of the A-E's who designed the existing facilities were only vaguely aware that data terminals would be in use. Even if they were aware of it, from their viewpoint there was little difference between data and the office telephone. Data processing professionals have been equally at fault for not pursuing these details with their A-E's. This chapter should clearly demonstrate that designing a facility for data and installing a proper network can solve a tremendous number of problems.

A data comm facility in general now having been described, Chapter 5 singles out the communications area of the facility for further study. This area is central to the network and is intimately linked to the host DP system. For that reason, certain network and facility topics need further discussion to cover the special needs and problems encountered. The high density of the equipment creates problems associated with power distribution and housing. In turn, connecting all this equipment to the network creates problems due to high-density cabling. The network, originating in this area, has its own problems due to the large amount of wiring. These topics and more are discussed at the implementation level to show how many of the problems in the comm area can be lessened or eliminated.

Special data circuit implementation topics are the subject of Chapter 6. Every data comm operation has those special data circuits which have to be treated a little differently. Some circuits really don't need a modem, but it takes some massaging of the wiring to make a terminal work without one. When you get that solved, up pops a special case where you need another wire-pair for proper signal handshaking. Everyone seems to have a drawer full of adapters, without which some practical functions could not be implemented. Special cables also are difficult to avoid. These items are required in most normal situations. More yet are required in abnormal situations such as in electrical noise or lightning suppression problem areas. It's nice to have these special concoctions to get around certain problems and, in many cases, implement applications more economically and reliably.

Chapter 7 discusses some conventions and practices which should be followed by the people in the data comm shop. Most often, this is the area where circuit implementation becomes very sloppy; it is so easy to go ahead and do a job the way one wants to do it rather than follow some prescribed method. Few supervisors bother to check on how the practices are followed. Supervisors whose background is largely in software find this area difficult because wiring the physical network is not part of their experience. Data comm is an especially detailed area in which to work. There are ports, cables, adapters, circuits, power outlets, connector types, color codes, labels, and termination hardware by the score. Working with them in a conventional manner saves time and money and avoids frustration.

Chapter 8 gets us into an unpopular area, paperwork. It is hoped that the momentum gained up to Chapter 8 will carry us through this chapter, where our attention becomes easily distracted. Documentation is one of the necessary evils in the data comm business. To do our job, the documentation must be accurate and trustworthy. If it is not, there is little point in maintaining it. Because of the quantity of documentation required to keep track of the data communications function, the task is not simple, and it requires a dedicated effort by all concerned. Topics of interest here are the types of documentation needed to accomplish the work both efficiently and accurately. This includes documentation available from outside sources, such as journals, technical references, and books, and paperwork generated internally, such as processing forms, checklists, and line listings. Internal documentation also includes the inevitable policies and procedures, another necessary evil. I think most people will agree that, even though we don't like these topics, we cannot ignore them and still be effective.

Chapter 9 continues our endurance test of the less interesting topics and discusses the maintenance, personnel, and management functions in data communications support. Considerations which go into the maintenance program are discussed—primarily the pros and cons of doing it yourself or having the vendor do it. Personnel and the various levels required are considered only in quite general terms and only for common situations. Guidelines are included for the management function, which should prove useful to most shops.

Last, and especially in this case, not least, the Appendix provides checklists and aids for implementing some of the ideas presented in the text. This is intended as a reference section; therefore, information normally required to implement a circuit can be found here. Specialty tools, connectors, pins, wire, and the like are included for easy reference. No attempt has been made to provide a complete list of vendors. I have included the manufacturers who in my own experience have good products at reasonable costs. This is not to say that other manufacturers' products will not suffice as well.

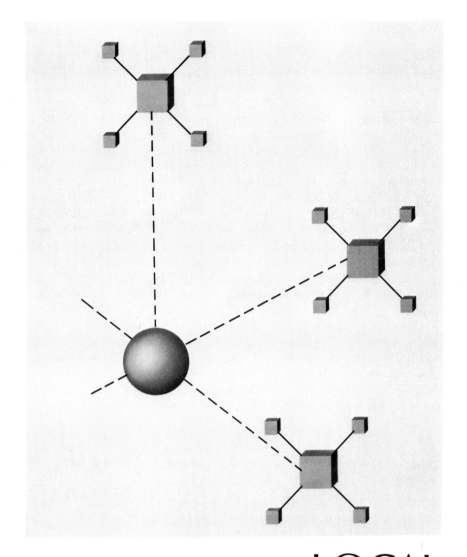

LOCAL NETWORKING

CHAPTER 2

14 LOCAL NETWORKING

Local networking is a broad area which covers many applications. Here we will be considering local networks in a limited sense. Very basically, our concern will be how to implement local networks for common data processing terminals connected to a DP service within a building or between several buildings in a campus situation. Further, the medium will be only twisted pair or 327X coaxial cable. Some would say that this limited application should not be classed or defined as a local network, primarily because a more sophisticated local network is not in use. This argument is countered by the very large number of facilities populated with computer terminals using twisted pair and coax cable.

In the past, people have not paid much attention to these simple networks they have created within their facilities. The networks grew with really little notice of the fact that they existed. No planning was done in the early stages to avoid problems which might occur in the future. This attitude spilled over into the facility design also. Not being aware of internal networks being created, people ignored the need for facilities to be designed with networking in mind. Inter- and intrabuilding communications is indeed a form of local networking, and a fairly large one at that.

Of concern in this chapter are the physical aspects of twisted pair and coaxial media. Protocols, electrical transmission detail, and the like are not of interest. Physical characteristics of the media will be of interest in connection with their incorporation into the facility environment. The scope of this chapter is limited to a conceptual level of facilities networking in order to lay the groundwork necessary for implementation details given in later chapters.

MEDIA

Figure 2-1 illustrates the types of media we will be concerned with throughout the text. Twisted pair cable comes in a wide variety of pairs and gauges. Of principal interest to facilities networking are 22 and 24 gauge wire in groups of 1, 2, 25, 50, 100, and 200 pairs. This range suffices to implement most every need. All network twisted pair cable except the two-pair will be solid conductor. Two-pair cable, used for terminal drops, is preferably stranded wire cable. All network cable of 25-pair or greater will be identical with that used in the telephone industry.

With as little technical detail as possible, it might be of interest to mention why twisted pair cable is used. Basically, a signal can be transmitted over wire media by using a signal wire and a return wire. Two conductors are needed to carry a signal reliably from one point to another. These conductors are almost always placed in an environment with surrounding electrical noise. The noise is coupled into the conductors and becomes part of the signal being transmitted. This noise, then, degrades the signal and introduces errors. The more unequally noise is coupled into one conductor with respect to the other, the worse the noise problem

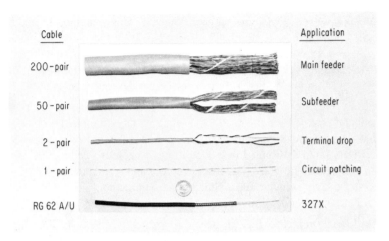

Fig. 2-1 Media examples.

becomes. When the two conductors are twisted together, external noise is more likely to be coupled equally into each conductor. On the receiving end, this equal noise injection can be effectively canceled out, leaving only the signal. This is a low-cost solution to reducing the noise susceptibility of the cable as compared to the use of two parallel conductors. Twisting wire only lessens the noise problems, it does not eliminate them.

The use of coaxial cable allows implementation of higher-performance data circuits. There is basically only one type of coax of concern here. It is RG 62A/U, which is now in widespread use for implementing 327X circuits. The reason it is used is twofold: it supports a wider communication bandwidth required for 327X applications, and it is less subject to noise than paired cable. Again being quite simplistic, coax reduces noise by the way it is physically constructed. The center conductor is the signal wire. It is completely surrounded by a braid, which becomes the second conductor for the circuit. Surrounding the signal wire makes it much more difficult to couple noise into the signal, and so the performance of the cable is improved.

Even though twisted pair and 327X coax are the only interest here, it would perhaps be of value to mention other media in use with data communications systems:

- The performance of twisted pair cable can be improved by surrounding the pairs with a conductive wrap that is similar to the braid of a coaxial cable. This is known as a shielded twisted pair cable. The technique is commonly used to improve performance in systems designed to communicate over twisted pair. It should be noted that the extra shield makes it difficult, but not impossible, to use this cable in existing twisted pair networks with common telephone "punch blocks" for terminating paired cable.

- Ethernet-type coaxial cable networks are implemented by using baseband communication with a special coaxial cable channel. This channel is shared by all devices connected to it. Baseband refers to sending the "raw" digital signal directly into the coax rather than modulating it with a carrier signal first. The coax itself is considerably different in size, performance, and cost from that found in 327X applications. Terminals and other devices connect to the Ethernet via special interface boxes physically close to the main channel cable.
- Broadband coax is used in situations where a variety of signal types are transmitted over one coax channel. Here the data signals are mixed with TV and other signals and distributed over a building or campus. The main channel is again a special type coax, larger and more expensive than 327X coax, and generally a rigid, tubular coax. Small coax, similar to 327X coax, is tapped from the main channel to the individual terminal locations.
- "Fiber" media refers to glass or plastic filaments used to carry light between points. The light is modulated or pulsed to transmit information and has a capacity for carrying very large amounts of information. Use is presently most cost-effective in very high data rate applications, but it is becoming more common at terminal data rates. Multiplexing equipment using fiber media is available, and it can be installed in a facility to reduce both wiring and noise susceptibility.
- "Carrier current" refers to devices which use the facility power line to carry information between points; this is very similar to wireless intercoms which simply plug into the wall outlet. Since this method is in quite limited use for data and requires no network as such, it is of no consequence here.

It should be pointed out that all these forms of media, except carrier current, influence the facility design detail for networking. The techniques described later permit integration of these less common media, as well as twisted pair and 327X coax, into the facility.

The media of a network can be configured in various topologies, some of which are shown in Figure 2-2. In many respects, the twisted pair networks to be implemented here will resemble an unrooted tree topology. This is from the network viewpoint, in that the network can be used to connect any device to any other device effectively. From the terminal viewpoint, the network is used in more of a point-to-point fashion, in that a terminal is simply connected to a host. From a host viewpoint, if the network is used to connect all terminals to the host, the topology looks like a rooted tree or star. For large interbuilding local twisted pair networks, multipoint circuit topologies can be formed. Twisted pair media, then, has many network topologies, which are somewhat dependent upon one's viewpoint and the context of the application. Coax, on the other hand, is more restricted. Present 327X terminals using coax have a central controller. Terminals can either be configured in a point-to-point fashion or be daisy-chained with some

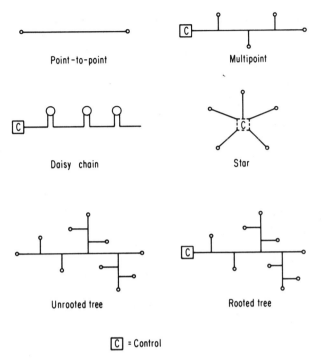

Fig. 2-2 Common network topologies.

equipment. From the controller's viewpoint, a star topology would exist. These definitions are useful in viewing the networking from various perspectives.

Media can be used in various ways within the network. Coax is quite restricted; it connects directly to the terminal at one end and the controller at the other, with little variation. Twisted pair networks, however, can have several flavors. Modems can be attached to modulate the data before it is shipped over the network. This can be done point to point with simple line driver modems (point-to-point topology) or in a multipoint fashion with multiplexing modems (multipoint topology). In other cases, terminals can be connected directly to the host front end without modems. Intercoms and current loop devices may also add to the variety. This mixture does have an impact upon the network. Signals over the network can range from low-amplitude modulated carriers to rather high-amplitude baseband digital transmission. This variety needs to be considered in the network design.

A look into the crystal ball is necessary to complete this section on media. What is in the future for twisted pair and coax? A look at terminals and their applications provides some clues. Interactive productivity is very important in many applications. Screen devices (327X) rather than line devices (async CRT) increase productivity considerably in many instances. This more than offsets the higher cost and assures significant use of coax well into the future; twisted pair simply cannot supply the bandwidth necessary for some of these applications.

On the other hand, there are still many applications with smaller bandwidth requirements. Asynchronous CRT terminals are still in large demand and will continue to be popular because of their low price and utility for general applications. There is no point in spending more money for both the terminal and media if the higher performance cannot be taken advantage of by the application. Technology will also continue to advance in the types of devices connected with twisted pair cable. The much lower cost of the twisted pair media will continue to give reason for this development. The telephone company has a large investment in twisted pair and will continue to use that media for data in the foreseeable future. In twisted pair facility networks, the network can also be used for connecting intercom systems, word processing systems, point-of-sale equipment and terminals, etc.

As RS-422, -423, and -449 gradually replace RS-232, an increase in twisted pair utilization should be noticed. Presently, much of the bandwidth of local twisted pair cable is unused. RS-422 [10 Mbaud to 4000 ft (1200 m)] and RS-423 [300 Kbaud to 2000 ft (600 m)] will allow for more effective use of the available bandwidth. Indeed, the distances in the newer EIA specifications are ideal for twisted pair local networking.

Another factor affecting the future of terminal usage is the "paperless office" concept. It is true that DP systems will continue to increase the number of office terminals; therefore, a corresponding increase in communications media will be required, and much of it will be twisted pair. All things considered, there is no reason to expect a sudden death of twisted pair communication media. In summary, it is quite worthwhile to design and implement local networks into facilities with both coax and twisted pair in mind. The future will require it.

IMPLEMENTATION CONCEPTS

Emphasizing one last time, we are concerned with networks using twisted pair or 327X coax as their media. The concepts to be developed will allow other media to be used effectively in DP facilities as well, but they will not be discussed in any detail. Knowing a little about media now, we need to think about how to design the cable layout to form a facility network. The actual mechanics will be discussed in Chapter 4; for now, we will stay at the conceptual level. For network implementation, consideration must be given to the components used and to the physical conditions within which the network is installed. Before getting into the details, let's first look at some of the problems we are trying to overcome:

- Why should networks evolve by stringing one wire at a time as new terminals are required? With a little forethought the facility and installed network can be designed to eliminate this problem. This will be a principal goal.

- We should form an understandable network. Typical networks, having evolved a circuit at a time, are almost incomprehensible because of the many types of wire in use, the junctions and splices in obscure places, and the physical difficulty in even manually tracing the path that the circuit follows.
- An understandable network will be documentable. It is very difficult to document the *real* network in many existing facilities. It's easy to say that a circuit leaves Port A of the system and goes to room B of the building. That is not adequate documentation; the points in between also need to be documented.
- A network, once installed, should be easy to maintain and change. This requires some attention to the physical design of the facility.
- The network and facility must be able to function in a dynamic environment. Things do not remain static in data processing; change is routine. Technology is always advancing, requiring changes in our established ways of doing things. Media, as already mentioned, change, and that will cause new problems for facility networks in the future. Buildings should be so designed that all this can be taken in stride as much as possible.

Twisted Pair Network Concepts

Solving facility network problems requires us to concentrate upon the concept of a wired facility. The task at hand is to install a network within the facility, first somewhat ignoring what will be attached to it. The network should be like an octopus with tentacles reaching to all points within the facility, possibly even to other buildings. Convenient network access points should be at the ends of the tentacles to allow connections to various equipment. Since the network generally exists to support a terminal population connected to a host, the network should terminate in a central wiring panel in the data communications area, close to the host data processing support equipment. Here access to the telephone data circuits also is possible.

The resulting network is illustrated in Figure 2-3. Notice the boundaries of the facility network. The network definition is not dependent upon the type of equipment attached to it. Network access points are also the network demarcation points between the network proper and the equipment using the network as a communication path. The network consists of the central wiring panel, the network access points, and all connecting cable in between. Links to other building wiring panels can be considered as part of the same network or tie points to separate networks.

For a better idea of what the network components look like, refer to Figures 2-4, 2-5, and 2-6. Figure 2-4 illustrates a small central wiring panel. It could also serve as a remote wiring panel connected to a larger central wiring system, as shown in Figure 2-5. Figure 2-6 shows a terminal block which serves as the network access point. Blocks of that type are distributed throughout the facility; they connect to the central wiring panel via main and subfeeder twisted pair

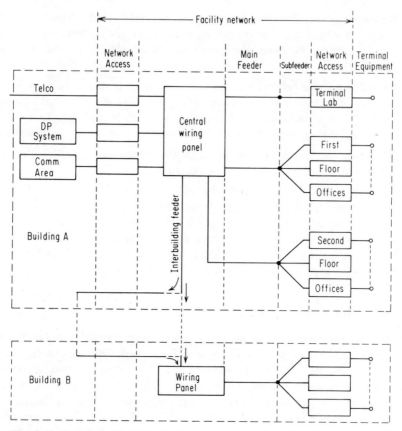

Fig. 2-3 Conceptual twisted pair network.

cabling. Two types of terminal blocks will be discussed later: one for terminating solid conductor cable and the other for terminating both solid and stranded conductor cable.

Forming a network as depicted in Figure 2-3 involves the idea of access zones. The facility needs to be broken down into various zones into which network access points are installed. Shown in Figure 2-3 are several zones: the DP system, the communications area, a terminal lab, and the basic floors of the building. As shown, each floor is further broken down into three other zones. Every facility would have its own unique number and types of zones. A zone would generally be considered the area served by one network access point or perhaps a few points.

The terminal block shown in Figure 2-6 allows the termination of 50 pairs from devices accessing the network. A typical terminal, for example, would require a send pair and a receive pair to implement the circuit. In this case then, one terminal block could provide service to 25 terminals in its zone. Further, if

Fig. 2-4 Small wiring panel system.

only terminals were connected to it, the terminal block could provide service to 25 offices, each with an associated terminal. By taking into consideration the applications required—terminals, intercoms, WP systems, etc.—and other factors such as potential growth of these applications, zones can be established within the facility.

Why should a twisted pair network be implemented as shown in Figure 2-3? The reasons are many:

- The solution uses commonly available components. The punch block is a widely used and reliable method of terminating twisted pair cable. This is the same technology that the telephone company uses.

- The cabling is commonly available Telco-type twisted wire cable. Since Telco uses this cable in huge quantities, the cost is low and a large number of sources exist.

- The wiring techniques are easily understood, and specialized tools exist to make mass termination easy.

- Running large feeder cables to various zones eliminates stringing one wire at a time. Labor costs are high compared to material cost. It's not much more

Fig. 2-5 Large central wiring panel system.

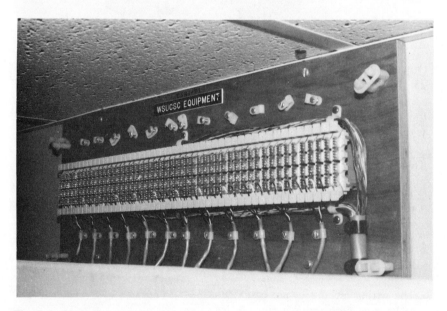

Fig. 2-6 Network access block.

difficult to string one larger cable which can serve 25 or more terminals than it is to string one small cable for only one terminal.
- Flexibility is unprecedented. The central wiring panel allows for many application variations, circuit patching schemes, quick patches, and normal network changes.
- The network can grow as needed. Only the required sections, and not the whole network, must be installed. As zones become saturated, it is relatively easy to add another terminal block for additional network access. The central wiring panel system also can grow with little disruption, as will be seen later.

Interbuilding networking is simply an extension of the same concepts as used in the facility. The same components, cable, tools, and techniques are used. The main difference is that the cable paths may be more restricted between buildings than within a building. Conduits are often required between buildings to implement the network. This requires that proper planning take place to determine both the number of pairs required and the amount of spare conduit needed for future growth. Transient suppression also becomes a concern between buildings, but as explained in a later chapter, it is easily solved.

Coaxial Cable Network Concepts

Implementing a wired facility by using coaxial cable is a little more difficult than by using twisted pair cable. Compromises have to be made because of the additional cost of the media, the larger size, and the difficulty of forming large feeder cables out of individual coax. Since coax is more restricted in the possible application topologies, coaxial networking does not have the generality that twisted pair networks have. The basic requirements are illustrated in Figure 2-7. In systems with multiple 327X controllers, it may be desirable to have coax patching equipment, as shown in Figure 2-8, to allow easy alteration of the network. A feeder cable will consist of 8, 16, or more individual coax bundled together to form one larger cable. The idea behind using feeders is to ease the task of stringing cable into major areas or zones of a building.

For example, if five terminals were required on the third floor, with the 327X controller on the first floor, it would be advisable to install a feeder consisting of eight coax between floors. If a major growth were expected on this floor, the feeder size should perhaps consist of 16 individual coax. The feeder is installed from the controller area just to the entry point of the third floor. Since the feeder installation is probably the most difficult, the cost of the additional cable in the feeder is traded for labor costs in doing this job several times. Once feeder access is provided to the third floor, it is relatively easy to run individual coax between the third floor entry point and the individual terminals throughout the third floor. In a like manner, feeder cable can be installed to various other zones within the

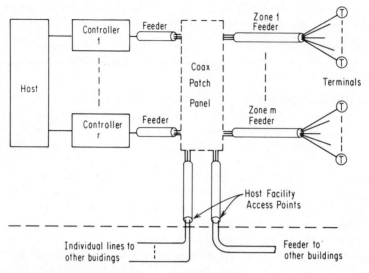

Fig. 2-7 Conceptual coax network.

Fig. 2-8 Coaxial cable patch panel.

facility. The reasons for doing so are the same: it's less costly in the long run; it allows more rapid connection of new circuits; and it's more flexible.

For a twisted pair network it was possible to formulate a fairly precise definition of the extent and boundary of the network. For coax networks, it is more difficult to be that precise, and no attempt will be made here to define the extent and boundary of the coaxial facility network. Expanding the coaxial cable network to an interbuilding basis is again an extension of the same principles as for twisted pair. Physical access to other facilities is probably the most difficult problem.

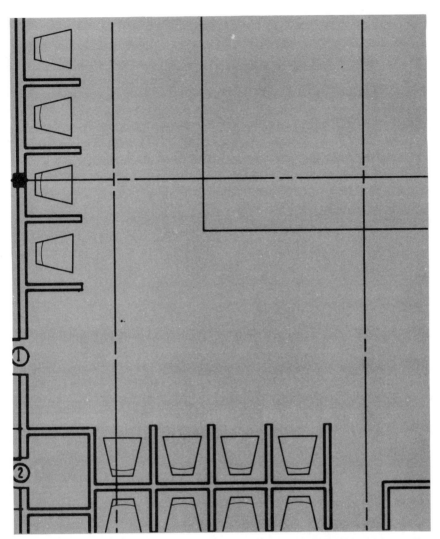

FACILITY DESIGN

CHAPTER 3

28 FACILITY DESIGN

A few years ago, DP facility design efforts were always directed to the main machine room and its special needs such as electric power, raised floor, and cooling. Then this area generally encompassed the entire equipment base including the data communication network function. The design phase of the main machine room usually was done by the architect-engineer in cooperation with the computer professionals.

The results have been generally satisfactory for the machine room and its operation. Design of the remainder of the facility, however, would typically be accomplished as for any other commercial building. Invariably, the special problems related to housing remote terminals and other distributed processing equipment in the same or adjacent facilities were ignored. Properly designed cable paths into the hallways, offices, and between floors are a luxury few installations have today. Office design to integrate the telephone, terminal, and hardwired communication lines also was remiss.

At a campus level, interbuilding communication design considerations at the facility level were nonexistent. The reason for this poor design was, and still is, not so much the fault of the A-E; instead, it was due more to the process by which facilities become designed. The intermediaries who often are involved between the A-E and the end user filter out important facilities-related information. The A-E is generally not in a position to suggest design strategies specific to data processing for lack of the necessary specialized knowledge. The end user, although a computer professional, typically is not versed in the various networking options and technical information to be an effective interface to the A-E. Additionally, end users generally have their own jobs to do and don't want to be bothered with the seemingly endless details involved in the design and construction of the data processing facility.

Today, in many cases, the above comments still apply. Facilities in which the needs of distributed processing are ignored still are being designed. In other cases, people have recognized that DP-intensive facilities need special consideration throughout the facility, not just the computer room. The end of the 1970s seems to be a transition period when people became more aware that terminals were going to proliferate in the future. That's putting it mildly; an explosion of use would have been more accurate. What we want to develop now, then, are some facility implementation strategies to solve some of the problems. This book as a source of ideas needs further explanation. Throughout the text, many of the concepts presented are illustrated with actual working examples. The intent is to demonstrate that the ideas are practical. No suggestion is made that the schemes shown are the only way to do it. The designer still has some freedom to try other ideas; the important thing is not to lose sight of the special facility needs. With this philosophy in mind, some of the problems we will be considering are the following:

- What considerations need to be given to office design? How do you design data cabling into offices, keeping in mind that people like to move their furniture? What about data cable, telephone, and power interrelationships?
- How do you design the hallway areas to allow cabling to reach the individual rooms?
- How do you design the vertical access space between floors?
- What special considerations need to be given to high-density service areas such as terminal labs?
- How do you design the facility, knowing that data requirements will undoubtedly grow and change as a matter of routine, and yet still retain the flexibility needed?

The ideal, of course, is to design a new facility with all the features necessary to implement practical networking within the structure. There is no reason why the discussions here need be limited to new facilities, however; many of the ideas to be presented can be modified to work in existing ones. It's certainly more difficult after the fact, but not impossible. Depending upon the need, it may be entirely cost-effective to remodel an existing facility. Each situation needs to be judged independently. Consider the ideas presented here to be applicable to both new and old facilities.

TERMINAL-END CONSIDERATIONS

We will consider the aspects of facility design from the viewpoint of the end user and work back to the communications area, where the data service originates. The end users want their computer service available as conveniently as possible. This means in their offices, not in someone else's office or away from their places of work. Facility design should be carried out with this in mind and be adjusted to the particular situation. If the entire facility will be data-intensive, every office should have access to the network within the structure. It would be a mistake to design 90 percent of the facility with access to the network just to save a few short-term dollars. Facilities change almost as soon as they are completed. The best-laid plans are soon modified when the facility becomes occupied. Needs change; people change; goals change. It is only too soon discovered that the 10 percent of the offices originally left out of the data design also need access to the network. On a different level, perhaps only certain areas of the facility, such as a particular floor, are data-intensive. This floor, then, should be designed for total access to the data services. As a rule of thumb, provide as much access to the network as possible; the fact that it is required at all should be warning that much more will be required in the future.

Office Design Considerations

Concentrating now on the office itself, we should consider:

- How will data cabling enter the office area?
- How many places should it enter?
- What about placement with respect to the telephone?
- What about placement with respect to power?
- How do we design for people moving their furniture around?

There are a few facts we can base decisions on. For data, we know that computer access will be from either a hardwired line or via telephone access. In either case, power is required for a terminal. Additionally, short-haul modems or dial modems may exist, and they also will require power. To meet these requirements, the hardwired line, the telephone jack, and the power outlet must be within close proximity to one another. To solve the problem of where, an obvious solution is to provide data, telephone, and power access on all walls of the office. This is not practical—it is too costly, unless it can be justified by special circumstances.

The other extreme is to provide service access on only one wall. That limits flexibility. A reasonable compromise is needed. If the budget permits, the compromise would be to install the basic services on two opposing walls in each office. This probably means extra cost for both the telephone and data service outlets. Power is no doubt already provided on two opposing walls for some of the same reasons we are trying to provide for data. In regard to power, it is generally safe to assume we are considering standard outlets. There will be times when 208-V three-phase power will be required, but they should be considered special cases and not the rule.

Figure 3-1 shows the basic services required in the office. The triad of power, telephone, and data should be duplicated on an opposing wall if at all possible. As shown, the data service is provided via a standard electrical box with, at minimum, a TV-type outlet plate installed. Outlet plates are available which have coax or RS-232 connectors installed. These are useful in certain situations. The data outlet leads back to the hallway just outside the office via conduit installed inside the wall. This allows data cable to enter the office neatly. In contrast, it is not uncommon to see cable entering through a hole in the ceiling and hanging down to the terminal.

Figure 3-2 shows how the conduit is terminated into the hallway; here, it comes out into a hallway cable tray for ease in cabling. A plastic bushing should be installed in the conduit to prevent the cable from being cut as it is installed.

Implementing offices with services on opposing walls will allow the cabling to be easily installed into the office and with at least an alternate route to conform to the location of the terminal in the office. This allows the cable to be neatly

FACILITY DESIGN **31**

Fig. 3-1 Basic office services for data.

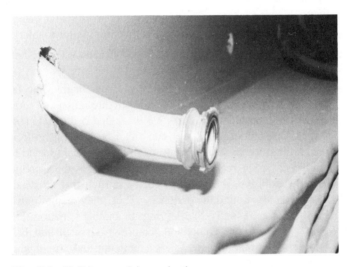

Fig. 3-2 Hallway conduit termination.

dressed and put out of the way. It not only is a safety feature but also keeps the cable out of the way of feet and movable furniture, which tend to damage the cable over a period of time. The conduit will readily take either twisted wire cable or coax. As a matter of fact, several cables of each type could be installed in the same conduit if the situation demanded.

As a note of caution, make sure Telco uses its own conduit and not the data conduit. It can be disastrous to mix Telco with your own cable. In the future you may find the need for extra space to pull more or larger cable. Telco also may need to pull a larger cable. Keeping those potential needs in separate conduits can spare you a lot of hassles. Under no circumstances can the data cable be installed in the same conduit as the power cable. These are the reasons for three separate services—power, data, and telephone—each in its own conduit leading out of the office.

Terminal Labs, Equipment Rooms

When the situation calls for a room, such as a terminal lab, to have a high density of equipment, the design requirements change. It is much more difficult to provide standards to follow because of the variety of situations possible. The room may or may not have a false ceiling and/or raised flooring. The furniture may be built-in or movable. The basic concerns are:

- How is the data cabling routed into the room?
- How is power provided to the room?
- How are power and data distributed within the room to all the equipment needing them?
- How do you provide power and data while also providing for safety?
- What are some of the user concerns that should influence the design?

Raised-floor rooms are perhaps the easiest place to begin. The false floor allows both power and data cabling to be distributed throughout the room, which simplifies many of the problems. Figures 3-3 through 3-5 illustrate some of the possibilities with raised flooring. Power can either be brought up from outlets installed beneath the floor or designed into the room as usual (providing for the increased number of outlets required). The choice depends upon circumstances. In Figure 3-3, for example, the room geometry is such that terminals can be practically installed only around the periphery. Thus, they require power outlets only on the walls and none below the flooring. In another situation, Figure 3-4, a larger room with a mix of equipment requires both wall and below-floor receptacles. The last example, Figure 3-5, is a more unusual case in which raised flooring was tiered for a classroom situation. Here, because of the layout and the furniture in use, raceway is used on the furniture to carry both the terminal power and the data cables coming from below the floor. Data cable will undoubtedly always come up from beneath the raised floor in these situations. The main data feeder providing service to the room must enter the room and be terminated somehow. The floor offers the most flexible alternative.

Therefore, in raised-floor situations, be sure the design includes a way to install the data feeder beneath the flooring. The problem here is that the feeder will probably come into the room via the hallway overhead space. This will require that a vertical drop exist to get from the hallway to the floor of the terminal lab. For power, check for a conduit path back to the electrical service panel for that area. It may be wise to design in an extra empty conduit or two for the installation of special power which may be required in the future. Special power can easily be accommodated beneath the flooring if it is planned for. Other design strategies for raised-floor areas will be considered later, in the communications area design. Some of the ideas presented there may also be applicable to terminal lab situations.

FACILITY DESIGN   33

Fig. 3-3 Small raised-floor terminal lab.

Fig. 3-4 Multi-type terminal raised-floor lab.

Standard rooms without raised floor can tax one's ability to find good solutions to the cabling problems. Raceways installed around the periphery of the room are perhaps the best solution. The furniture, being movable, can be laid out to conform to the raceway. This will allow the data and power cabling to originate from a wall and permit using the furniture as a path to the terminal equipment sitting on the furniture as shown in Figure 3-6(a). At least this approach has some

34 FACILITY DESIGN

Fig. 3-5 Tiered raised-floor terminal lab.

flexibility. Other solutions are to design the facility with outlets installed directly in the floor as shown in Figure 3-6(*b*). For this to work, one must be very sure of the layout. Even then, as needs and equipment change, there is no way to relocate the fixed floor outlets.

Rooms full of equipment are typically high-population areas. The design of the room must take this into consideration to have the approval of fire and other safety people. There are good reasons for this. Generally, rooms of this type may house 30 or so devices and possibly an equal number of people. Designing the facility for two egress paths would be prudent. Check with the proper safety people to resolve any doubt. For emergency situations, the safety people also frown on data or power cables hanging where a person can become entangled in them. Consider these factors in the facility design. More information on ways to implement safe rooms will be presented in a later chapter, where the details of the network cabling installation are given.

HALLWAY CABLING REQUIREMENTS

Generally speaking, a data cable will enter the hallway of a building through one or perhaps a few specific locations. This applies to in-house circuits as well as telephone company data circuits. The problem is to design the facility to allow network cabling to travel from the hallway (or floor) entry point to the individual offices and other rooms needing access to the network. Most facilities will only allow two choices: run the cable either above the ceiling tiles or just below the tiles on the hallway walls. Sometimes a mix of the choices will be in use. In any

FACILITY DESIGN 35

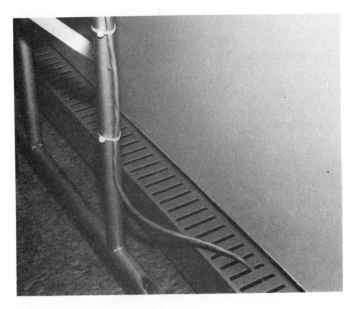

(a)

(b)

Fig. 3-6 Standard floor cabling. (a) Peripheral raceway; (b) floor outlet.

case, a cable tray or raceway should be used to contain the cabling. Design goals would allow cable trays or raceways to traverse both sides of the hallways, have access to any room off the hallway, and have crossover points from one side of the hallway to the other.

A good design method from the network viewpoint is shown in Figures 3-7 through 3-9, which are views of cable trays, open on top, built into the hallways just below the ceiling tiles. This allows rapid and very easy installation of cabling throughout the building. It also adds a different touch to the architecture. To be seen in the photos are the crossovers for routing cable from one side to the other. They should be designed with inside corners made of a material which will withstand the force of cable being pulled against it. Gypsum wallboard, for example, won't take the abuse. Conduits to the individual rooms and offices should terminate in the cable trays as shown in Figures 3-10 and 3-2. This can apply to both data and telephone conduits. The scheme allows for very easy cable routing from the hallway trays into each room requiring service.

Cable trays like those shown in Figures 3-6 to 3-10 obviously cost more than other solutions in the short term. In the long term they save a lot of time, giving an economic payback. Other things can be done to offset the cost; the most practical one is to use the cable trays for the telephone wiring along with the network cabling. That will save considerable money in conduit and cable installation labor charges, which will in part offset the tray costs. Any other signal wiring, such as intercoms, could also be routed through the trays. Most building codes would allow any type of signal cable to be installed in the trays but not power wiring, unless it is enclosed in metal conduit.

Another solution to running cable in the hallways is shown in Figure 3-11.

Fig. 3-7 Hallway cable trays.

FACILITY DESIGN   **37**

Fig. 3-8 Hallway cable trays.

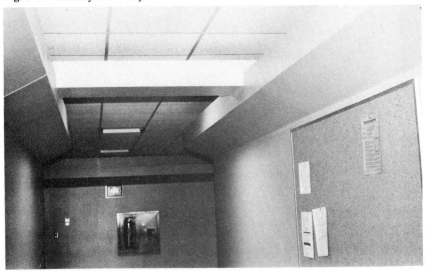

Fig. 3-9 Hallway cable trays.

Here the raceway is installed above the ceiling tiles. Although not as convenient as trays below the ceiling tiles, this is a good method for installing cable trays. Figure 3-12(a) shows a way of cabling a hallway where fixed ceiling tiles are in use. Conduit of this type is more difficult to use than open cable tray, but certainly it is easier than installing cable above the fixed ceiling tiles. Figure 3-12(b) shows how to install a terminal drop using the raceway in Figure 3-12(a). Install the outlet box in the office with a small conduit through the wall connected to the

38 FACILITY DESIGN

Fig. 3-10 Conduits entering cable tray.

Fig. 3-11 Above-ceiling cable tray.

(a)

(b)

Fig. 3-12 Below-ceiling cable raceway. (a) Hallway raceway; (b) terminal drop installed from raceway on opposite side of wall.

40 FACILITY DESIGN

hallway raceway. This scheme for getting cable into a room from the hallway is particularly useful in buildings with rigid ceiling tiles. In some cases, there may be little choice but to run the cable above the ceiling without any conduit, tray, or raceway at all.

VERTICAL CABLE ACCESS

Vertical cable runs are required to get from floor to floor or from the hallway overhead to areas under any raised flooring. One characteristic of vertical runs is that the cable making the run is likely to be a large feeder rather than an individual terminal drop wire. The design should take this into consideration and allow for a larger space. Also, large feeder cables are more difficult to bend around corners; this too should be allowed for in the design. Cable weight is still another problem. Cables should not be bent over a corner with the full weight of the cable supported by the corner bend area. These points are better made by Figures 3-13 through 3-16:

- Figure 3-13 shows a building support column which has been firred out to provide a vertical cable run space. Small doors in the column make the cable installation easier.

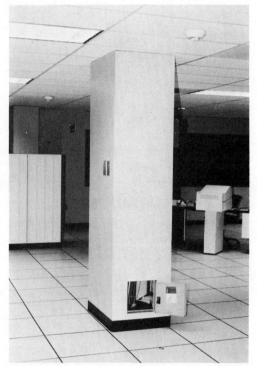

Fig. 3-13 Vertical cable run firred into support column.

(a)

(b)

Fig. 3-14 Access doors for vertical cabling. (a) Located near ceiling; (b) opening into hallway cable tray on other side of wall.

42 FACILITY DESIGN

Fig. 3-15 Vertical cable strain relief.

- Small doors can also be installed in walls as shown in Figure 3-14. In the case of Figure 3-14(*b*), the door has been designed to open into the cable tray on the opposite side of the wall. This makes it quite convenient to install cable feeders into a floor area.
- When long vertical runs are made, it may be necessary to install strain reliefs. Figure 3-15 shows a piece of plywood installed on a wall where cable clamps can be used to help support the cable weight.
- Figure 3-16 shows a solution to the problem of coming up through a concrete floor. Typically, a hole is left in the floor through which cable can be routed. The sharp edge of the original hole is eliminated by molding concrete around the hole. This reduces potential damage to the cable.

INTERBUILDING CONSIDERATIONS

Most large facilities have service entrance points for the basic needs of the building such as water, power, and telephone. This area would also be used to run cable for extending the local network from one building to another. In other

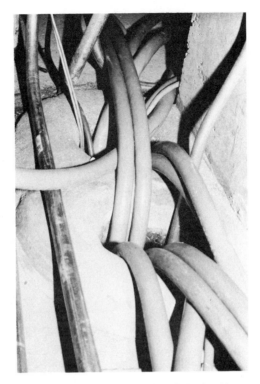

Fig. 3-16 Molded corners for vertical cable runs.

instances, excavation may be needed to install underground conduits. No matter how the routing is accomplished, some thought should be given to the needs of the network when it leaves the facility. It is always wise to install a termination board where cable leaves the facility.

An example of a termination board is shown in Figure 3-17. The board should have space for installation of punch blocks, hanger space for coax, and a little extra space for both growth and the installation of special things such as transient-suppression devices. The details are given in a later chapter. For now we are concerned only with the facility design aspects. The design should also allow some space for service loops. It is good practice to leave extra cable near the termination point just in case it is needed in the future. Notice both the service loop and the need for cable hanger space in this area. The service loop can be invaluable when an error is made in terminating the cable and just a few more inches of wire are needed at the other end or when the cable has been damaged maliciously or otherwise.

In the placement of the termination board, consideration should be given to foot traffic patterns, security, and potential water damage. Service entrance points are typically in difficult places and are frequented by people who are used

44 FACILITY DESIGN

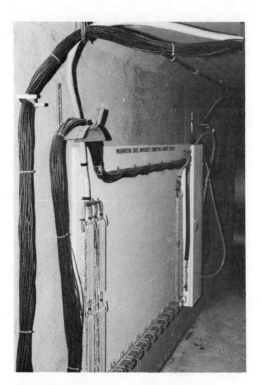

Fig. 3-17 Interfacility termination panel.

to working with sledge hammers, not delicate wiring. Keep these points in mind in the design.

Interbuilding networking frequently involves conduits to install the cable in. The conduit can present serious problems if not properly handled. Consider the following in your design:

- Slope a conduit in such a way that water cannot be trapped anywhere within the run. Also, don't forget to consider the lower end; allow a place for the water to fall without damage to any equipment in the area.
- Where a conduit must change direction, make as gradual a curve as possible. Try to keep the curve radius over 3 ft (1 m). The preferable material for the bend area is metal, because plastic conduit is easily damaged by pulling cable over a curved section of it. In addition to greater durability, metal conduit has the advantage of less friction when cable is pulled through it.
- When initially loading cable into conduit, it is best to fill the conduit to capacity. Pulling wires one at a time soon causes problems. The wires become tangled and intertwined with the pullwire to the point that no further cable can be installed even though the conduit is not filled. The wires themselves can also become damaged from the strain of pulling when they are entangled.

FACILITY DESIGN   **45**

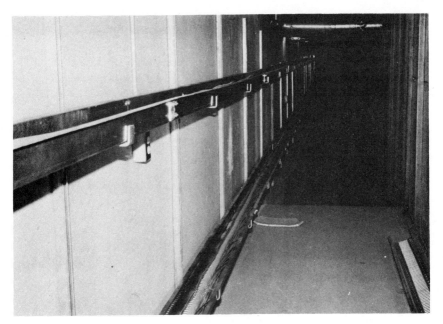

Fig. 3-18   Open cable tray.

- It is preferable to install many smaller conduits than one large one. Smaller conduits allow smaller quantities of cable to be installed initially, and a conduit can be filled immediately at less cost.
- The ends of the conduit should be dressed to prevent damage to the cable being pulled into them.
- As soon as possible after the conduit is installed, put permanent labels at each end to clearly show the length of the conduit. Use a stencil and spray paint. This will prevent a lot of frustration later on when you need to order cable but cannot remember how long the conduit is.
- Pullwire should be given serious consideration. If the conduit is all metal, no particular problems should occur. Plastic conduit, however, has inherent problems. For plastic it is advisable to use as large a pullwire as possible to prevent slicing the conduit as cable is being pulled. Large plastic marine rope is ideal for this application.
- Open cable tray, as shown in Figure 3-18, is useful in certain applications. It can easily be attached to service tunnel walls in multiple runs to help segregate the various types of cable.

COMMUNICATIONS AREA CONSIDERATIONS

Up to this point, discussions have centered on distributing the facility network throughout the building and to other buildings. The source of all the network cabling is the communications area. All the hallway trays, vertical runs, conduits, and other paths lead back to the comm area. Here the network originates with all the support equipment necessary to implement the service—Telco, front ends, modems, diagnostic equipment, the network circuit patching panels, etc. Because of the density of the equipment and the large number of circuit cables, the comm area is considered in detail in its own chapter. In this facility design chapter, we are more interested in the gross features of the facility necessary to support a network. In this light, we need only make a general comment about the comm area: design it as part of the main computer complex and its associated needs, including raised flooring, protected power, Halon, special power provisions, and a secure environment.

INTEGRATING THE NETWORK AND FACILITY

CHAPTER 4

At this point, we have an idea of a local network and also an idea of a facility designed with networking in mind. We now need to link the two ideas together to produce a facility designed and implemented for data communications. In this chapter we will consider all of the facility network areas except the communications area. Again, the comm area is special enough to warrant a chapter of its own to handle all the details. Our goal now is to go into as much detail as necessary to show how a network is implemented in practice. Before that, however, we need to clean up some details not included in the earlier network discussion.

FEEDER CABLE CONCEPTS AND PRACTICES

Twisted Pair

Feeder cables have been mentioned in general terms up to now, but more detailed information is required to provide an understanding of why and how feeder cable is used in a facility. Twisted pair feeders of interest here will be paired in groups of 25, 50, 100, and 200. This cable will be used to meet the bulk wiring needs of the facility. Refer again to Figures 2-1 and 2-3 to see the area of interest. As a guideline, cable is generally used as follows:

- *Main feeder cable* is defined roughly as the cable providing service to a major area of the facility, such as a wing or floor. Several main feeders may be needed to supply the major area. A 100- or 200-pair cable would most often be used for this purpose. In some cases, a 50-pair cable might be classed as a main feeder, as when servicing a cluster of terminals in one room. Another application would be to use several 50-pair cables in parallel in place of, say, one 200-pair cable. This is sometimes desirable to reduce signal coupling between pairs in long runs. The closer pairs are to one another, the more one signal can be coupled into an adjacent pair. Using multiple 50-pair cables in place of one larger feeder puts some distance between many of the pairs and coupling is thereby reduced, at least between the individual 50-pair cables.

- *Subfeeders* are classed as cables providing service from the main feeder to individual zones within, say, a floor of the facility. Since most zones are terminated by a 50-pair network access wiring block, 50-pair cables would be the most common subfeeders. Sometimes each half of a wiring block is connected via 25-pair cable.

The characteristics of all feeder cable are identical. The cable is exactly the same as that commonly used by the telephone company. It has the familiar gray or tan jacket, and the pairs inside are color-coded with the five major and five minor color-coding groups. See Chapter 7 for color-coding details. For implementing facility networks, 24 gauge wire is recommended here for most purposes for the following reasons:

- It is a compromise size and is readily available. Telco uses wire ranging from 22 to 26 gauge in data applications. The 24 gauge size falls between those two sizes.
- Punch blocks are designed to use 22 to 26 gauge wire. Using the larger 22 gauge tends to spread the block clip a little, so that, if a 26 gauge is used later, the connection is not as tight. Using 24 gauge is a compromise in this situation.
- Ideally, 22 gauge would be used because its larger size permits longer data runs and it is a less delicate wire to work with. On the other hand, 26 gauge would be the worst as far as size yet best as far as cost is concerned. Again, 24 gauge is a compromise choice.
- Mass termination connectors and tools are frequently used to terminate the feeder cabling. The 22 gauge size is a little on the large side for certain connectors and tools; 24 gauge is a better physical size.
- Terminal drops can be either solid or stranded cable, which requires that blocks that terminate zones be able to utilize wire of either type. These special blocks more easily accommodate 24 gauge wire with its smaller diameter.

Coaxial Cable

Coaxial feeders, as required here, do not exist—they must be made. One simply lays several individual coax cables in parallel and then ties them together with cable ties spaced every 2 to 3 ft (60 to 90 cm). The ties should be just tight enough to hold the cables together yet allow a little slippage when bending the bundle around a corner. There is no coax parallel to twisted pair as far as main feeder versus subfeeder goes. In practice, coax feeders will comprise 8, 16, or 32 cables bundled together. One might consider a 32-cable bundle as a main feeder and the others as subfeeders, but it is difficult to form any useful definition.

Refer again to Figure 2-7 to visualize the coax feeder requirements. Why these sizes? There are at least three reasons: cost, size, and controller compatibility. Groups of 8, 16, and 32 adapt to 327X controllers nicely where these particular numbers of coax can form major and minor groups. In certain cases, one might consider a group of four coax as a feeder, but labor versus material cost must be considered in making a feeder out of fewer coax. Feeders also have to be limited in size the other way. Too large a feeder can be overly costly for the application if many of the individual coax within the feeder will never be used. Feeders larger than 32 coax also become difficult to manipulate physically. Everything considered, 8, 16, and 32 bundles work well and can be used in a cost-effective manner. The following guidelines should be used for coaxial feeders:

- Short feeders should be built in common multiples, the shortest being around 25 ft (7 m). Standard sizes of 25, 50, 100, and perhaps 150 ft would be appropriate. (Metric lengths of 10, 20, 30, and 50 m would be reasonably standard equivalents.) These feeders are commonly used to connect controllers to patch panels or nearby terminal labs. Never custom-cut the feeder to fit the particular requirements at the time. Equipment has a habit of being moved around. Rooms also change from their intended original purpose. Custom cables become troublesome in this changing environment. Even if the application requires a cable just slightly longer than one of your standard lengths, use the next longer standard length. The extra length is not much more costly, and it can pay off in the future when someone decides to move equipment. Short feeders can also be joined with barrel connectors to form longer feeders if necessary.
- Long feeders linking major areas of the facility together can be custom-cut. To save costs, connectors need be installed on individual coax within the feeder only as required. Unused coax can always have a connector installed later when a new circuit is needed.
- Anytime two coax are joined with connectors, insulate the connection. Large plastic tubing commonly used for chemical lab applications works quite well as shown in Figure 4-1. Never leave the metallic connectors exposed where they could come in contact with other circuits or even metal parts of the facility; 327X equipment may not run properly if this happens. Figure 4-1(a) shows a typical splice where one would be tempted to omit the insulation, i.e., there is only one splice in the whole feeder. But without the insulation, the metal coax connector will physically contact the metal cable tray and possibly cause some problems. In Figure 4-1(b), two large coax bundles are connected; notice the use of the service loop concept.

BULK FEEDER CABLING INSTALLATION

It is assumed that, prior to installing the bulk feeder wiring which forms the facility network, zones have been established and rough cable routes chosen to optimize time and cost. After that, it is time to actually lay the feeder. Since the facility design has in some sense predetermined how the feeder cabling will be installed, there is not much more to say except to list a few guidelines:

- Make gradual bends, not sharp ones.
- In vertical runs, don't hang the full weight of the cable over a corner. Use a strain relief somewhere; refer again to Figures 3-15 and 3-16.

Fig. 4-1 Insulating coax splices. (*a*) Splice in a coaxial cable run; (*b*) coax feeder connection point.

52 INTEGRATING THE NETWORK AND FACILITY

- Be careful not to cut the cable jacket when installing the cable.
- Try not to parallel power wiring.
- Avoid laying cable close to large motors or other equipment which uses large amounts of electric current.
- Watch out for foot traffic areas. Don't install where people will be walking on the feeder.
- Leave a little extra cable either as a service loop or just simply extra long. This can pay off immensely if a mistake is made.

If the feeder cable is being installed in a facility with open or accessible cable trays or raceways, the job is fairly simple. Cabling will be difficult in older buildings, especially ones without removable ceiling tiles in the hallways. About all you can do here is grin and bear it. Just make sure to pull enough feeder the first time.

Main Feeder to Subfeeder Linkage

After laying it, the bulk cabling is terminated into either a connector or a wiring block. Main feeders need connectors installed so that they can be joined to subfeeders. In turn, the subfeeders will terminate into the wiring blocks in the various zones. Figure 4-2 shows a 200-pair main feeder connected to four 50-pair subfeeders. The work is done efficiently by using standard connectors and the termination tools listed in the Appendix. The gender of the connector is determined by two arbitrary rules. The central wiring panel (discussed in Chapter 5)

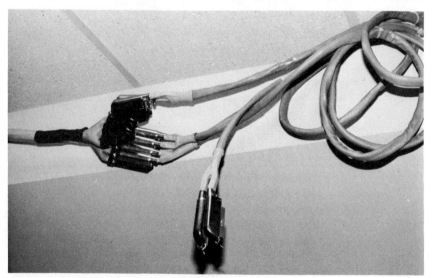

Fig. 4-2 Twisted pair main to subfeeder linkage.

has female connectors, which means that feeders connecting to it must have male plugs at that end. That's the first rule. The second rule is to use standard extension cables with opposite genders at the ends. Considering a feeder as an extension cable would require that it terminate into a female connector at the far end from the central wiring panel. Continuing with the subfeeder, the end connected to the main feeder would then have to be terminated with a male connector. At the block end of the subfeeder, the cable is terminated directly on the block and there is no need for a connector.

Figure 4-3 shows the parallel situation for main feeder to subfeeder linkage using coax. Note the use of the plastic insulation sleeve. Coax cable is always terminated with special coaxial connectors and joined together with barrel connectors. This ensures that the proper impedance of the cable is maintained. Trying to splice coax without the use of connectors is risky and, as a general rule, should not be attempted. Tools and components for terminating coaxial cable are listed in the Appendix.

Terminal Wiring Blocks

Terminating the feeder cable into wiring blocks is shown in Figure 4-4. Two cases exist: blocks to which only solid conductor cable can be connected and blocks which can terminate both solid and stranded wire. This distinction is from the terminal drop standpoint, not from the feeder cable's, because the feeder is always solid conductor cable. The blocks shown can be fabricated on-site or

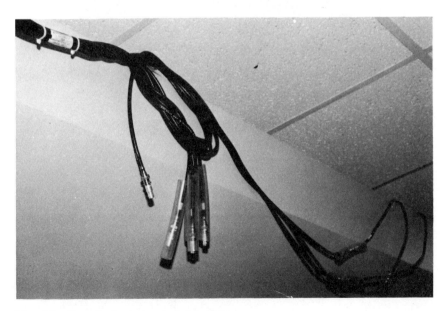

Fig. 4-3 Coax feeder to terminal drop linkage.

(a)

(b)

Fig. 4-4 Network access terminal block types. (a) Solid wire termination block; (b) stranded wire termination block.

purchased, depending upon the need. They should be clearly labeled as to their ownership. Being scattered throughout the facility, it is easy for Telco people to think that a block belongs to them.

Figure 4-5 shows a terminal block located just over an open cable tray where all wiring is very accessible. Note that the feeder does not directly route to the block panel; it first forms a service loop to leave a little extra cable before it is terminated. Hints on wiring the terminal blocks are given later when wiring and other practices are discussed.

Interbuilding Topics

Running feeder cable between facilities will probably be done by routing through either service tunnels or dedicated conduit. In service tunnels, some useful guidelines are the following:

- Keep cable away from power lines and steam lines. Position them where they are least likely to be damaged by the normal activities which take place in service tunnels.

Fig. 4-5 Hallway terminal access block.

- Position cable with water in mind. That is, don't place it where it will lie in water or be exposed periodically to water as below a manhole cover.
- Use a cable hanger with a wide surface area [0.5 in (1 cm) or so] upon which the weight of the cable can rest as shown in Figure 4-6. Here a rubber strip was added to help protect the cabling. Hanging by a small wire may eventually result in damage to the feeder insulation.
- When splicing cables, leave a service loop as shown in Figure 4-7. In this case the service loop would be more appropriately called a drip loop because it should carry any water away from the splice point.

When installing feeder cable in dedicated conduits between facilities, keep these points in mind:

- If 22 gauge wire is ever to be used, this would be the place for two reasons. Since the typical run is long, using a larger wire would result in less signal loss. The second reason is that the cable is stronger and less likely to be damaged by the pull through the conduit. However, 24 gauge will work well even in these applications.
- Use cable grease or talc, if necessary, to make the pull easier and place less stress on the cable.
- Use a wire pulling "sock" to pull the cable. This is a device similar to the novelty item which you put on the end of your finger and can't pull off.

Fig. 4-6 Feeder hanger detail.

INTEGRATING THE NETWORK AND FACILITY   57

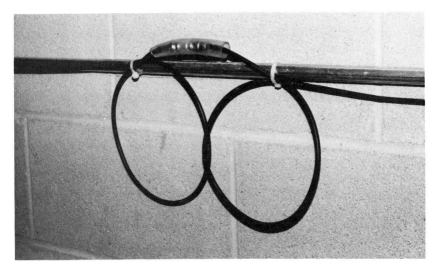

Fig. 4-7 Service/drip loop detail.

- Load the conduit as full as possible without risking damage to the cable. You can't get it much fuller than 75 percent or so because you need room for attaching the pull sock.
- Make sure the end of the conduit has a bushing installed to prevent damage to the cable as it is being pulled.
- To get a pullwire into an empty conduit, use a small wire or twine connected to a plug of cloth. Put the plug into the conduit and then use a vacuum cleaner to blow the plug to the other end. (There are also commercial devices to do this.) Once the small wire is installed, it can be used to pull a larger pullwire through.
- If you are using a power puller, be sure to pull slowly if you are using plastic conduit. Otherwise, the heat buildup from pulling over bends in the pipe can cause damage to the conduit.

As mentioned earlier, it is sometimes desirable to run several smaller feeders rather than one larger one to help reduce signal crosstalk. Long runs between facilities may be another place where smaller feeders are appropriate.

TERMINAL-END TOPICS

It is not uncommon to notice terminals wired with solid conductor cable. Part of the problem in doing this is that solid conductor cable is so noticeable. It doesn't flex, and it becomes an eyesore in a short time. Also, because of its inflexibility,

58 INTEGRATING THE NETWORK AND FACILITY

it can develop shear stress problems within the copper wire and eventually break after enough bending. The reason for the use of solid conductor cable is nebulous; it's hard to find a good reason. Today a variety of components and tools are available to use stranded wire cable efficiently and cost-effectively.

As a rule, this book recommends stranded wire cable for the terminal drop. In particular, stranded twisted pair cable of 22 gauge can be used very effectively. There is no reason not to use 22 gauge stranded wire rather than the 24 gauge recommended for the network wiring. Twisted pair should be used for the total length of the circuit from the comm area to the terminal. At times it is tempting to use nontwisted cable at the terminal end, but the temptation should be resisted. Maintaining twisted pair throughout the total length of a circuit keeps the electrical impedance relatively constant. Also, the signal-to-noise ratio is higher.

Insisting upon the use of stranded wire cable for the terminal drop creates the requirement that most of the terminal access blocks of the network be able to terminate stranded wire. For this reason, the majority of the terminal blocks should be of the type shown in Figure 4-4(b). They are considerably more costly, but they are well worth the extra expense.

Office Wiring

The terminal drop cable starts at the terminal wiring block and, ideally, leads to the office via conduit and comes out into the room to the terminal. Figure 4-8 shows a stranded twisted pair cable emerging from the wall conduit box. The cover plate has been removed to show the implementation. The knot behind the plate serves as a strain relief for times when the cable in the office is accidentally pulled. Figure 4-9 shows several cables coming through the same conduit into the room. In such a case, a tie-wrap can be installed around the cables behind the plate to serve as a strain relief. Figure 4-10 summarizes the typical office drop by showing the cabling coiled up near the terminal to keep it away from feet and

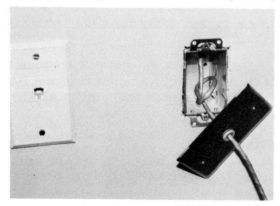

Fig. 4-8 Terminal drop strain relief.

Fig. 4-9 Multiple terminal drops in one conduit.

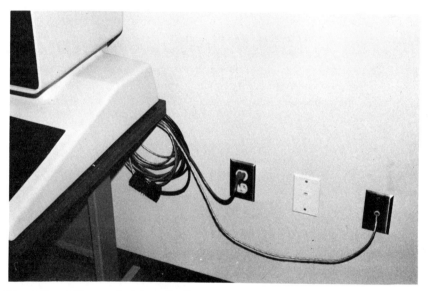

Fig. 4-10 Office terminal drop detail.

other things which may tend to harm it. Maintain a usable standard length for the office drop cable. If a conduit is available on opposing sides of an office, for instance, use a length of the drop cable such that the cable can service its side of the office regardless of where the terminal is located on that side. Again, custom-cutting causes problems later on when someone wants the terminal located at the other end of the office. It's much better to use the little extra cable and have length available to suit people's habits of moving around.

Terminal Labs, Equipment Rooms

Other techniques are useful when wiring a room full of terminals or other distributed data processing equipment. Generally, we would assume that the data service feeder would enter the area and terminate in a terminal wiring block. In most cases it would be impractical to wire each device in the room back to some remote terminal block servicing that zone. There are times when this can happen, and most of the comments in the office wiring section would apply. The problem then is to wire each device to the block in the room. It would be preferable to have the block centrally located under a raised floor. Often this is not possible, however, and the suggestions made here will have to be modified for the particular situation. Again the first rule is to not custom-cut the cable. It is much more practical in the long run to go ahead and set a standard length for all cable drops from the service block to the separate terminals. That length would typically be the longest possible distance from the block to a terminal in the room. This especially applies to a raised-floor area, where it is so easy to move equipment around. In a standard room, where the cable must be placed in raceways around the periphery of the room, it may not be possible to use standard-length cables. Some compromise may then have to be made.

This section is summarized in a series of photos which show various implementation tactics. Figure 4-11 shows a nice way to handle the cabling for a table-mounted terminal. Also, refer again to Figure 3-5. Here the data and power cables share the same space until they travel under the floor. Mixing power and

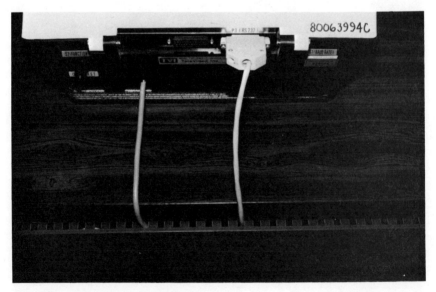

Fig. 4-11 Table-mounted cable raceway.

data cable in these situations will not cause any problems because the cables are quite short. Floor cable guards, Figure 4-12, are sometimes the only practical solution to the problem of installing terminals in the room interiors. Figure 4-13 is intended to demonstrate the flexibility of using cable raceway around the periphery of a room without raised flooring. Here, power cable, round signal cable, and ribbon signal cable are channeled via the raceway to their destinations. Figure 4-14 demonstrates something which is sometimes overlooked. In both

Fig. 4-12 Floor-mounted cable guard.

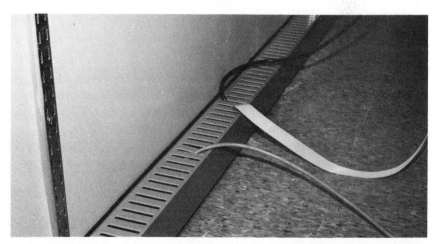

Fig. 4-13 Wall-mounted cable raceway.

photos the cabling is attached to the tables upon which terminals are placed. In Figure 4-14(a) the tables are movable a few inches (centimeters). Notice the service loop between tables to prevent cable damage from the person who gets mad at the terminal and tries to shove the table through the wall. In Figure 4-14(b) the tables are bolted to the floor so that the cabling can be rigidly attached to the table.

FACILITY AND NETWORK LABELING

So now you have a network installed in the facility. How do you refer to its parts and the path it takes? How do you document it? Labeling is required to allow the circuit to be documented. That doesn't mean you run around and put a few scribbles on things. Labeling must be done neatly to avoid confusion; it should be made consistent by following some standards; and it should be relatively permanent yet changeable. There are two things to worry about here. How do you describe the bulk wiring in the facility which encompasses many circuits within the feeder wiring, and how do you describe the individual circuit? The best answer is to use two separate documentations. It's too difficult and confusing to come up with a scheme to document both the bulk wiring and the individual circuit wiring. There are ways to short-circuit the documentation and come up with a scheme to handle both, but we are interested here in total documentation. It is not sufficient to document most of the network; all of it must be documented.

Bulk Feeder Labeling

What information is necessary to document and operate the building feeder portion of the network? The following items seem to be sufficient to do a proper job:

- An identification number is needed for reference in the documentation.
- It's nice to know where the other end is located. Remember that the feeder cable is relatively permanent and not likely to be moved around very much.
- Its ownership is sometimes valuable to know. There could be more than one network or area under separate control, which could cause a problem in regard to cable ownership. Ownership information also alerts Telco people that it's not their cable.

One advantage of feeder cable labeling is that the cable is large, which allows the use of a large label. In turn, a large label allows more information to be put on the label. This is generally good because it may save a few trips back to the book documenting the bulk wiring.

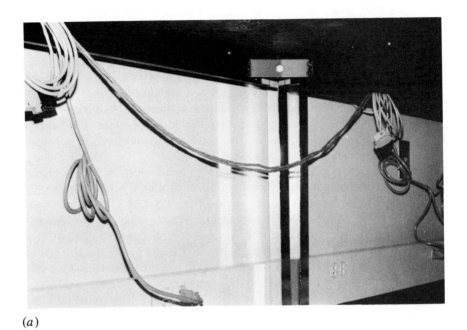

(a)

(b)

Fig. 4-14 Securing cable to furniture. (a) Horizontal cable run; (b) vertical cable run.

64 INTEGRATING THE NETWORK AND FACILITY

Figure 4-15 shows typical labels attached to both twisted pair and coaxial feeder cables. At times it is necessary to install a label in the middle of a feeder run, as when the feeder enters conduit to reach another part of the building. Since the cable cannot be physically traced because it is in the conduit, a label provides a convenient means of identification. All this leads up to the reason why an arrow is drawn on the cable label shown in Figure 4-15. When a label is in the middle of a cable, it may not be clear to which end the destination information applies. Therefore, when you put information about cable routing on the label, always include the destination of the cable. This is the natural way to do it, but labels in the middle of a run can get confusing.

Finally, the cable identification number must be assigned. Of the many ways to identify the cable, a good one is to use the format XXXYY-ZZZ, where XXX is the building identifier, YY is a subidentifier, and ZZZ is a specific cable identifier. For example, CSB2-007 would refer to the seventh feeder cable (007) installed into the second (2) floor of the Computing Sciences Building (CSB). Several variations of this scheme are apparent, and the format used should be tailored somewhat to the application at hand. For instance, four-place subidentifiers (YYYY) would be useless for small buildings. However, think globally about this. What if your network were to extend to a larger building in the future? It may be well to have more place positions in the label format for future growth. The same format can be used for either twisted pair or coaxial feeder cables.

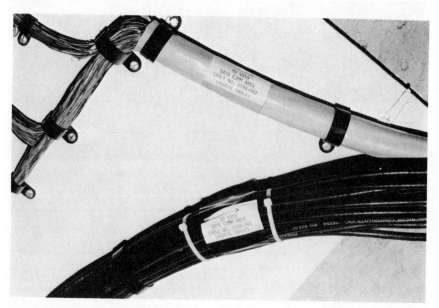

Fig. 4-15 Feeder cable labeling.

Terminal Block Labeling

Each of the terminal wiring blocks spread throughout the facility needs a unique identification number. The numbering scheme can be nearly identical with the labeling of the facility feeder cabling: XXXYY-ZZZ, where XXX is a building identifier, YY is a building subidentifier, and ZZZ is a specific block identifier. The format can vary a little to suit the needs of the particular facility. Figure 4-16 shows a typical terminal block and its identification labeling. Here the subidentifier refers to a block within a network zone.

The individual tab labels on the wiring blocks, also shown in Figure 4-16, are required to install the paired wires and document the circuit. The particulars of labeling the tabs are discussed later, when standards and conventions are considered.

Circuit Labeling

When working with the production network, it is the individual circuit which is of paramount importance, not the feeder cable through which the circuit routes. Day-to-day activities in using, changing, and troubleshooting the network involve the send and receive pairs which implement the various circuits. Feeder information is, for the most part, excess baggage to carry along once the network is installed in the facility. This is the reason for splitting the documentation into separate references for the bulk facility wiring and the circuit wiring. Once the network is installed, the information in the bulk wiring book is rarely used for normal network operations. That is not to say that the bulk wiring information is not important; it is important for other reasons which are explained in Chapter 8.

Fig. 4-16 Terminal access block labeling.

Circuit documentation requires that a label be installed at every terminal drop location using the network. The individual coaxial cables in feeders also may require unique identification. Figure 4-17 shows how both a twisted pair terminal drop cable and a cable in a coaxial feeder can be labeled with a circuit identification number. The twisted pair drop uses LDC (local data circuit) as a prefix followed by a four-digit alphanumeric suffix which specifies the individual circuit. For the coax cable, CDC stands for coaxial data circuit, and a similar suffix is used for the specific circuit. There is no particular reason why a separate prefix, that is, CDC or LDC, must be used. Each network can have its own scheme to best suit the situation. Normally, a circuit is given a number and a label is attached to the terminal drop. The circuit is then connected to the terminal. The circuit information is entered into the active network documentation, which shows the details about the circuit such as the type of terminal connected and the port to which the circuit is connected.

All this is fairly standard across installations. Now, what do you do about documenting the circuit when the terminal is to be disconnected? This is one area which is often overlooked in circuit documentation. When removing a terminal, it is best to leave the drop installed for use if it is needed again in the future. It's not worthwhile to remove it. After a network has been in use a few months or perhaps a couple years, a collection of unused terminal drops will form. You just can't forget about them, and you can't remember them either. They should be documented.

One way to deal with this problem is to have documentation for inactive circuits. There are many approaches, and there may be no one best way. In the examples shown in Figure 4-17 an alphanumeric suffix is used to denote active

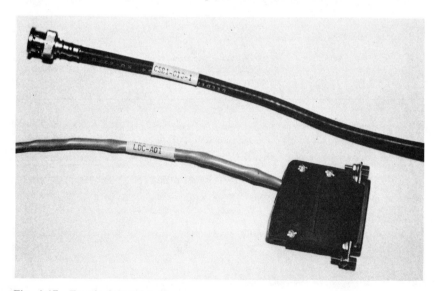

Fig. 4-17 Terminal drop labeling.

circuits. In another list for this network the inactive circuits are documented by using a purely numeric suffix with the same prefix (LDC or CDC). It's more important to document the inactive circuits somehow than to quibble about how to do it.

Interfacility Labeling Considerations

When the network leaves the building and forms an interfacility network, the building access points should be well labeled to help prevent problems. Building access points are frequented by Telco personnel, who may not be data trained. Confusion can develop when they see cable in use that looks exactly like some of their cable. Labels will make clear who owns the cable and has responsibility for it. Refer again to Figure 3-17, where an example of a building access point is shown. Large lettering on the access panel clearly informs everyone who has the responsibility for the cabling.

The rules to follow for labeling the cable in the access areas are the same as those already presented for general intrafacility cabling. Repeating an earlier suggestion, it is wise to label conduits as to their end-to-end length as shown in Figure 4-18. This avoids some frustration later on when people forget how long the conduit was when it was originally installed. It's a little difficult to measure cable length after installation.

Labeling Materials

Reviewing the labeling suggestions which have been made, it is apparent that basically two types of labels are required to label all aspects of the network. One

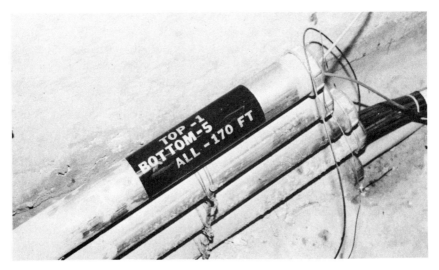

Fig. 4-18 Conduit length labeling.

is a cable label to wrap around the various main feeders, subfeeders, and drop cables. The other is general alphanumeric lettering for labeling the terminal wiring panels, access panels, etc.

The best cable labels to use are the long strip type which have an adhesive backing and a white area upon which to print the label information. They are also available in sheets for easy typing. During installation, the white patch area goes on first and is then overwrapped by the clear trailer portion of the label. This protects the printed material and forms a reasonably permanent label which can be removed easily if necessary. Once you get over the initial shock of the cost, you will find that the labels are worth the high price. The Appendix provides reference information for their acquisition. For general alphanumeric lettering, vinyl peel-off numbers and letters are available in a variety of styles and sizes from your local stationer.

WHY ALL THE NETWORK AND FACILITY BOTHER?

Except for the comm area, we have now pretty much covered how to design a facility for data and how to install a manageable and practical network within the facility. Now might be a good time to step away from all the detail that has been necessary and look at what we have accomplished. There must be some reason for going to all this trouble, and we don't want to lose track of it. Let's review some of the original problems and consider now what the facility and network can do toward making our jobs easier:

- We have an understandable facility and network. Both have been implemented by using reasonably standard practices, which makes both the facility and network consistent. There are few special cases, if any, to muddy the waters. Standard implementation allows for consistent documentation. Not only that, but everything is documented. New personnel have a reference for learning the facility and network; they don't have to rely on other people's fallible memories or waste their own time trying to figure out how a circuit runs. That is a valuable asset. Most of us know from our own experiences how difficult it is to operate without proper standards and documentation. It's not practical anymore to do things the old way. It's too costly.
- Both the facility and network perform. The facility allows changes to the network with reasonable ease. This is important in a field where change is the norm; terminals and other equipment are always on the move. The network has known cable characteristics and performance. If one terminal works without a modem, there is a very good probability that any number of terminals will work. You can count on the wire performance, in contrast to facilities which have a garden variety of cable types and performance is difficult to know or

control. The network can be used for a wide variety of applications with virtually no changes to the network itself.
- The network is cost-effective. The additional cost of implementing the facility and network will be more than repaid over the years from the time saved by people who are just normally working with the network. If ulcers have a price tag, the savings are immense. Errors are reduced considerably. The documentation task is eased. The network design takes advantage of commonly used cable, connectors, and blocks, which are economical to use because of their high-volume usage. Circuit implementation time is reduced tremendously; the terminal drop wire installation is the largest effort. All these things add up to cost savings and make the initial effort well worthwhile.
- The network and facility can function into the future. The facility, being designed to have paths throughout its extent, doesn't depend upon what is installed in those paths. In the future, you may want to install another network such as an Ethernet or a network using a fiber cable. The facility will allow you to do so.
- Don't forget the network users. Their offices are designed for data. They can even move their furniture around without daily tripping over their terminal drop cable. The network is more reliable, which shows up in the user's confidence in the system.
- And don't forget the network staff. Seeing them grumbling and cussing under their breath doesn't instill much confidence in the network users. If the network and facility are properly designed, people are encouraged to take a little more interest and pride in their work. That is another important asset, even though it is an intangible one.

COMMUNICATIONS AREA IMPLEMENTATION TOPICS

CHAPTER 5

The communications area as defined here is where the network originates. It houses the majority of the equipment base necessary to provide the data communications services: front ends, data comm diagnostic center, modems, Telco equipment and lines, 327X controllers, wiring/patching panels, etc. What is included in the comm areas of different installations will vary somewhat, but that will have no appreciable impact on our discussion here. This chapter has two goals:

1. To supply the previously omitted details of the facility comm area
2. To concentrate on features within the comm area which solve working problems and increase reliability, productivity, and efficiency

Certain characteristics of the comm area are the source of many of the day-to-day problems working in this area:

- A large variety of equipment introduces traffic pattern problems if the layout of the equipment is not done properly. Circuits being implemented with a string of serial elements (front end, diagnostic center, modems, Telco) should suggest that an efficient comm area layout will aid the troubleshooting process.
- The variety of equipment also creates problems as to where to locate and work with it.
- A high density of power outlets is required to support all the equipment.
- A high density of cables is needed to connect the various pieces of equipment together.

COMMUNICATIONS AREA FACILITY DESIGN

Layout

The basic equipment layout of the comm area is fairly straightforward and should present no particular difficulties for someone who works with data. For new facilities, give the room design some consideration yourself; don't let the architect or some other intermediary do it. Your problems won't be understood, but you will have to live with the end result. Being responsible for room design will help you in your final task of laying out the equipment. The layout process is a function of the space available. For that reason it is difficult to give specific rules about arranging the comm area. It's something of a puzzle to solve, but not a difficult one. Some guidelines to follow are these:

- Consider how all the external cabling will enter the data comm area and where it will be terminated in that area. This includes both Telco cabling and your own local network cabling. If the front ends and 327X controllers are not in the data comm area, don't forget to consider how their cables will be included.

- Lay the equipment out in a functional manner. The front ends and the controllers should go closest to the host area to minimize controller-to-host cabling. The diagnostic center should be centralized in the comm area between the front ends and the circuit equipment such as modems and Telco termination equipment.
- Where will you put the documentation needed in the daily operation of the network?
- Suggest that plumbing not travel across the floor in the data area. The space is needed for the high cable density. Besides, a water leak with all that cable can be a source of headaches. If the host is water-cooled, can a water leak eventually reach the comm area?
- Zone the comm area as a separate Halon zone. Don't let two adjacent Halon zones bisect the comm area; it's difficult to string your underfloor cabling through a zone barrier.
- A raised floor has been assumed. Let's hope it is not shallow; 12 to 18 in (30 to 50 cm) is recommended. A good tile size is 2 ft (60 cm) square.
- You may want controls to subdue the light over the diagnostic equipment screens.
- Have the concrete floor sealed. This will make cleaning easier and will keep the concrete dust down.
- Speaking of dust, are ionization detectors being suggested for under the raised-floor area? Heat detectors are not sensitive to the dust which always gets stirred up under the data comm area. It's a little embarrassing to tell your customers that you had a Halon dump and system shutdown because someone pulled a data cable under the floor, kicked up some dust, and tripped an ionization detector.
- Keep the building telephone circuit terminations completely out of the data area. Have you ever witnessed a Telco *telephone* person with disconnect orders come into a data area, listen for tone on some circuits, think that the modem carrier heard is dial tone, and then proceed to rip out your data circuits? Your customers don't want to hear that explanation for their service outage.

Power Distribution

The equipment required to implement the data communications support requires a lot of power receptacles. Two problems consistently pop up in this regard: When the equipment base changes, it seems that you always need a power conduit moved, removed, or installed. That is costly; it's not easy to modify the power runs with all that cable under the floor. The other problem occurs when pulling a data cable which is entangled with a power cord. The power cord may become unplugged and shut down a rack of modems—and, in turn, some of your favorite customers. These problems have solutions. Inadvertent unplugging of power cords can mostly be avoided by using twist-lock power plugs and receptacles. Conduit changing can be remedied by installing a power grid when the facility is constructed. Figure 5-1(a) shows a typical twist-lock power connection. Notice also, in Figure 5-1(b), that tying the power cable down near to the receptacle will help decrease the chance of entanglement.

The power grid solution is a little more involved. First, the large majority of equipment in the data comm area requires standard 115-V power. Only the front ends or controllers may require special power. We can take advantage of that fact and install a matrix of power outlets under the comm area raised floor. Doing so is costly. But consider what can be spent in power modifications over the life of the area. A matrix of outlets may only be 50 percent utilized; but when equipment is added or moved, an outlet will be close by. Figure 5-1(a) shows the outlet box inserted in a conduit run which is a leg of the power matrix. To form the matrix, have outlets installed on a grid spacing of 3 to 4 ft (0.9 to 1.2 m). Several outlets can share one conduit run; thus, even though the outlets themselves are on a grid, the conduit isn't. The electrician doing the job will have to figure out the best conduit layout for the particular geometry of the area.

As long as we go to this length to do the power outlets right, we might as well do the breakers right too. Each outlet should be wired to its own breaker to increase the system reliability. The outlet grid should be wired back to an electrical service panel dedicated to that area alone, with enough breaker space to handle all the grid outlets plus the special power outlets also required plus some growth room. The breaker panel should have individual breakers labeled clearly as shown in Figure 5-2. Also, go back to Figure 5-1(a) and note that the outlet of the grid is clearly labeled as to the service panel breaker to which it is wired. This labeling is a good habit to get into. Most of us can recall situations in which the DP staff or the facility electrician inadvertently cut the power to a box because of poor labeling. That's a reliability problem which should not exist.

Some consideration also needs to be given to special power provisions in the comm area. This power would typically be three-phase, 208 V. There is a way of increasing the utility of power runs over the usual rigid EMT-type conduit: use flexible conduit as shown in Figure 5-3. It costs a little more, but you can move it around with equipment changes. Follow these hints in regard to using flex:

- Make it long enough to reach a wide area, not just where you happen to need it at the time of installation.
- Be sure to have the electrician use stranded wire, not solid copper. Solid copper will shear after repeated moving.
- Dedicate a breaker to each flex, the same as the 115-V power grid.
- Label both the receptacle and the breaker.

(a)

(b)

Fig. 5-1 Power wiring detail. (a) Twist-lock power components; (b) power cable dressing.

Fig. 5-2 Breaker panel labeling.

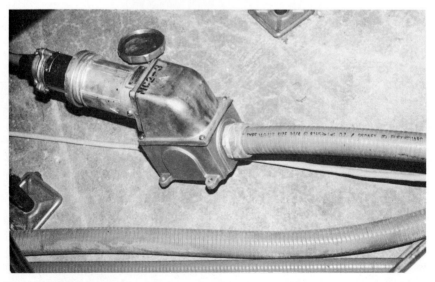

Fig. 5-3 Flexible conduit power run.

COMMUNICATIONS AREA IMPLEMENTATION TOPICS 77

Cable Distribution

What's the scenario in your shop when it comes time to make an EIA cable change in the comm area? When you lift the floor tile, do you feel you are opening a can of worms? Stepping into the mess, can you make it without damaging a cable? Will the mess let you get back out? When pulling out an old cable to relocate it, have you ever noticed how other cables looped around it can literally be worn down to bare copper? These are not fun things, considering the enormous number of cables which can amass in a data comm shop. If you have only a hundred or so cables, it may not be worthwhile to make any drastic changes. However, if you plan on growing considerably or already have a mess, you might consider an alternative: a cable grid similar in concept to the power grid. In what follows, the grid can be used for both EIA or coax cabling, or for any other cabling for that matter.

Figure 5-4 shows cable hangers specially built for data comm areas. To my knowledge, they are not commercially available and must be custom-made. Figure 5-5 shows the hanger components. Here is where the extra depth of a raised floor can be put to good use. Hanging the cables in this manner forces them to be installed neatly and with a little care. It also gets them out of the way of feet. They have their drawbacks too. They are costly to install initially, because several hundred hangers may be needed to provide adequate coverage. Also, their use makes cable alterations take longer because the cables must be manually placed in a hanger rather than just laid into the floor. The hangers do prevent

Fig. 5-4 Cable hanger detail.

78 COMMUNICATIONS AREA IMPLEMENTATION TOPICS

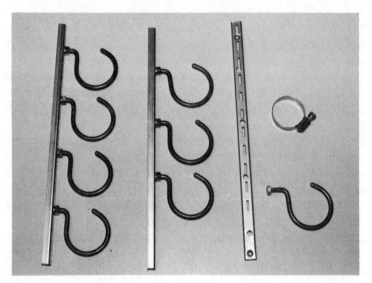

Fig. 5-5 Cable hanger components.

installed cables from being pulled out; they must be lifted out of the individual hangers. This stops cable rubbing and increases the reliability of the cabling.

The use of a cable hanger system as described here requires a commitment. It also requires a little more time and therefore increases network support costs. The initial cost amortized over several years should be low enough to be a minor decision point. The gains are twofold: a more reliable cable installation and a safer way to work with the cabling. The following suggestions should be useful if you decide to use such a cable management scheme:

- Install hangers on each vertical strut of the flooring system, don't skip every other one.
- In planning the hanger layout, consider intersection points where cables cross at 90 degrees. Figure 5-6(*a*) shows how. Assign one direction, say north/south, to always have hangers on one spacing, and assign the other direction, east/west, to have the hanger spacing shifted to allow cable to interleave in both directions.
- When making cable turns, you don't have to make abrupt corners; you can travel a 45-degree angle from one floor strut to its opposite corner as shown in Figure 5-6(*b*). This causes some underfloor areas to have cable in the standing space, but this is a relatively minor problem.
- Plan carefully what you will do with the excess cable; it must be stored somewhere. You might adopt the procedure, say, to coil up the excess EIA cable at the modem and dress it to length toward the front end.

(a)

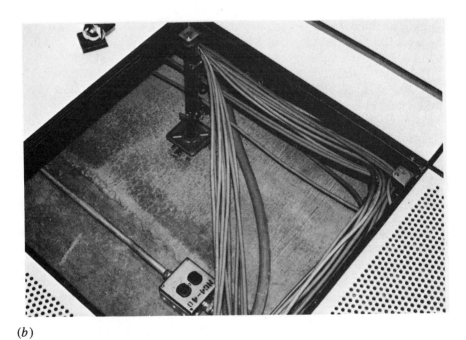

(b)

Fig. 5-6 Routing cable via hanger system. (a) Intersection point detail; (b) turning cable direction.

COMMUNICATIONS AREA FEATURES

Having considered the facility aspects related to equipment layout, power, and cabling, we now want to concentrate on some of the special equipment items which are installed in the comm area. In particular, we will discuss a scheme to terminate all the local network wiring at the comm area and also consider ideas for housing some of the various comm equipment. These particular topics seem to encompass a large number of the problems typically occurring in normal day-to-day operations. Few organizations have a system or procedure for managing their local network cabling. Typically, a terminal cable will come into the comm area and, more than likely, be connected under the floor to the communications controller. If a modem is in use, the cable will be forced up out of the floor where it becomes more manageable.

Telco twisted pair wiring is usually a separate entity within the comm area that has little physical integration into the overall networking function being performed. Without a physical networking scheme, this segregation is desirable. A little thought to the subject, however, can produce a systematic way to merge both Telco and your own networking needs into one consistent method of implementation and still retain some autonomy. The housing of the various data comm equipment is an interesting phenomenon to view among installations. One shop will adopt the operating room philosophy and place literally all minor support equipment (modems, etc.) in a cabinet and shut the door so it cannot be seen. The appearance is very clean. It's a bear to work with, however. At the other end of the spectrum is a spectacle that taxes your imagination that the network even functions. We will be looking for a compromise between these extreme situations.

Centralized Twisted Pair Wiring Area

Do you recall all that cable which was strung throughout the facility and leads back to the comm area? Now we have to do something about it. It must be terminated and organized in a fashion which is simple to work with and easy to document. Figures 2-3 and 2-5 demonstrated the solution to terminating the network which we will now go into more deeply. At the heart of the system are the large punch block wiring panels shown in Figure 2-5 and, in greater detail, in Figure 5-7. These wiring panels can be either custom-made by the user or purchased.

The basic idea is to put a large number of blocks on a panel, install drop cables on each block, and run the cables under the floor, where they are terminated with connectors. Feeder cables then connect to the wiring panel connectors as needed. By following a philosophy which forces virtually everything to be terminated at the wiring panel, any one circuit can be patched to any other circuit. A terminal on the second floor can be connected to the host communications controller as

(a)

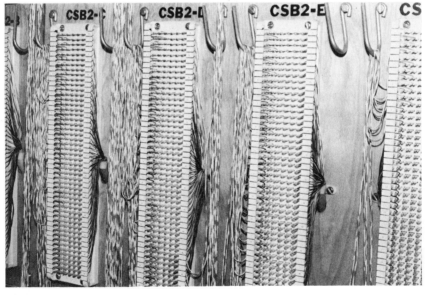

(b)

Fig. 5-7 Wiring panel detail. (a) Block organization; (b) block wiring detail.

easily as it can to a word processing machine in the next building. This generality forms an extremely flexible network.

The wiring panels are built in vertical sections 8 ft (2.4 m) high and 4 ft (1.2 m) wide by using standard sheets of plywood. Each section comprises four sections: a top panel, a bottom panel, and two wiring panels. The bottom panel is simply a space filler to force the wiring panels up off the floor to a convenient working height. The top panel is useful for labeling and also for cable access when the cabling is required to go up through the ceiling rather than to the floor. The wiring panels can be of a couple different varieties. Typically, each punch block has two attached 25-pair cables which lead back to the underfloor area.

If Telco personnel also use the central panel wiring system, they will need some bridge-clip blocks. These have a cable attached to the left side of the block to which the Telco circuit feeder can be connected. The right side of the block has no cable leading back out from the panel system, because wiring from that side leads only to other blocks on the panel system. Bridge clips are installed on the blocks to link the Telco data circuit to other blocks, as required to implement a data circuit. Bridge clips are required for circuit demarcation. Figure 5-8 shows a bridge-clipped block. By comparing the clipped block on the left to the non-clipped one on the right, you can see how the clips bridge one side of the block to the other. For those who desire more information on using a punch block wiring system, the Appendix contains information which should prove useful in dealing with some of the details.

Equipment Housing

Why do some people find it so unbearable to have cables come up out of the floor and connect to boxes for the whole world to see? I personally think they have seen a shop where the installing was done in a very unprofessional manner. It was probably a mess and not something the organization could take any pride in. Indeed, there must not have been much pride to start with to allow such a thing to occur. A lot of this strong feeling has occurred because the people who are responsible for network operations have little experience with the hardware side of the house and simply don't know how to handle the situation.

Look back at the panel wiring system. It's nice to look at, is very functional, and is easy to work with because of its accessibility. It's not buried somewhere out of sight; it's located right in the mainstream of the network activity where it should be. Even Telco can share the panel wiring system and clean up its act at the same time. How many Telco twisted wire termination areas which you have seen are cosmetically equal to the data comm area in which they are installed? Very few, I'll bet. Now I ask, why isn't the same philosophy used for all the boxes and components used to do networking? Why bury them? They should be accessible, not hidden away where they are difficult to troubleshoot or to cable.

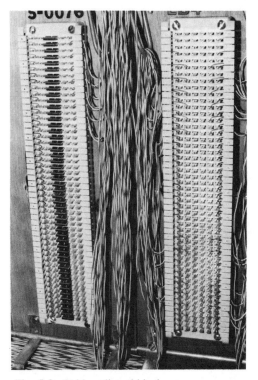

Fig. 5-8 Bridge-clipped block.

But how do you house all this stuff and still make it cosmetically acceptable? To some, you never will, but there is a solution to having a data installation that is both reasonably presentable and easy to work with. In Figures 5-9 through 5-11 are shown custom cabinets designed especially for solving some of the problems associated with housing data comm equipment. As viewed in the photographs, they are fairly self-explanatory as to their construction, function, and use. Some points to note are the following:

- The false bottom was designed both to get the lowest shelf up off the floor and to create a cable storage space accessible from the back. Getting the equipment up from the floor keeps it cleaner and therefore more reliable. The cable space is a useful place to put cable left over from dressing out the EIA circuit.
- The space at the top was similarly designed to keep the in-use equipment from getting too high and also to create storage space for spare equipment.
- The shelving can be moved to any desired height by using KV track (the vertical metal strips) or any similar hardware.
- The front of each shelf has a plastic strip in which to insert labels describing

84 COMMUNICATIONS AREA IMPLEMENTATION TOPICS

Fig. 5-9 Equipment shelving—front view.

the equipment and its place in the network. The strip lets you position a label at any point along the shelf.
- The back of each shelf has a spring custom-designed to dress data cable vertically along the back of the cabinet. This feature helps clean up the appearance of the shelving, keeps things in order for easy identification and alterations, and in general adds a different twist to working with data. You too may have your management wonder about your sanity when you budget request 50 to 100 custom springs for a network operation. This item may be a good test of how receptive your management is to doing networking in a practical manner in contrast to mimicking everyone else.
- The top back section allows space for installation of terminal blocks which are connected back to the central wiring patch panel. Here is another example of literally everything being wired back to the main wiring panel. The screw blocks shown provide a variety of terminations for the variety of equipment tered. Telco may scoff at the use of such outdated components, but they are cheaper and more practical than using a 50-pin connector for every data

COMMUNICATIONS AREA IMPLEMENTATION TOPICS 85

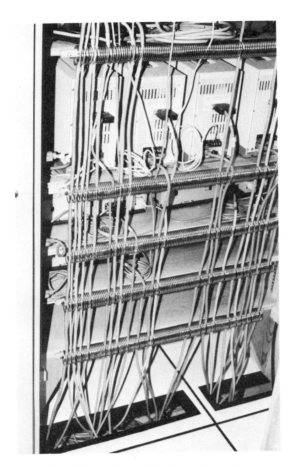

Fig. 5-10 Equipment shelving—back view.

circuit. This area can also be used for special terminations such as multiplexed (FDM, for example) circuits where resistive terminations may have to be added to the circuit.

- You need many power outlets. Power strips are very useful for this application, as shown.

Rack-mounted equipment also can be improved as shown in Figure 5-12. Very few equipment racks are set up properly as they are received from the factory or even from the vendor who installs equipment in the rack. Notice the improvements added to the commercially available racks:

- The same labeling strips as used on the custom cabinets can be used on racks to label the equipment attractively.

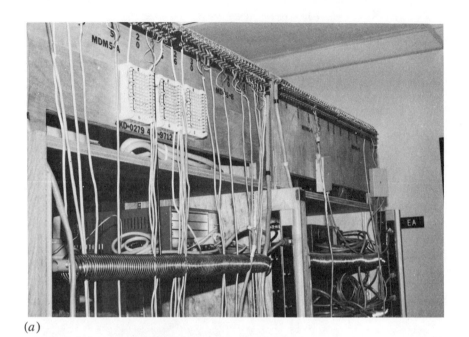
(a)

(b)

Fig. 5-11 Equipment shelving detail. (a) Rear top section detail; (b) cable spring detail.

(a)

(b)

Fig. 5-12 Modifying equipment racks. (a) Adding external labeling; (b) internal modifications.

- Other labeling can be attached to the rack for the referencing of components. This is useful for rapidly locating rack-mounted modems, for instance, where many may be in use.
- A power distribution strip adds flexibility to cabling.
- Twisted pair termination blocks are installed in the cabinet itself. This is part of the same philosophy: run everything back to the main wiring panel via a feeder cable.
- Notice the cable bar. It serves two purposes: some of the weight is removed from the connector assembly, and it serves as a tie-wrap fixture to dress the cabling neatly.

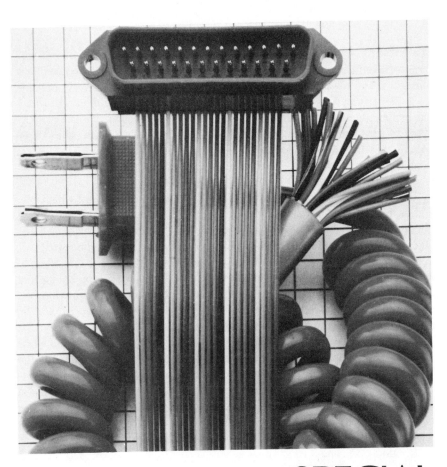

SPECIAL DATA CIRCUIT IMPLEMENTATION TOPICS

CHAPTER 6

Now to the fun stuff! Special data circuit implementation is how the network staff earns its bread and butter. By knowing networking at a fairly intimate level, you can alter it and perform functions more efficiently and cost-effectively in some cases. It's challenging to try to improve the service or to solve problems which the vendors don't seem willing to tackle. Who says you can't run data reliably over EIA runs longer than 50 ft without modems? Short-haul modem salespeople do! Are they making your decisions? If you have an installation with 200 async terminals, each with a pair of short-haul modems at a cost of $300 a circuit, who wouldn't want to sell $60,000 worth of equipment?

That is just one of many examples of an effort by the network staff which may both save money and increase reliability. One must be a little careful when getting into the custom gadget business, however. It's easy to get into the habit of doing quick and perhaps shoddy jobs and leaving no documentation trail on what was done. It has been my experience that if you cannot take the time to do the job in a professional manner and also take the time to document it, you should not attempt it. Otherwise, the end result rightly reflects upon the person who was responsible for it, and considerable confusion is generated among the people who are trying to support a custom device or modification.

There is no particular rhyme or reason to the flow of this chapter; the intent is to jump from one topic to another in areas likely to be of interest. Many of the ideas may not apply directly to your particular situation, but they may be adaptable to solve some problem in your own shop. Lastly, much of the discussion in this chapter involves EIA signals. The older RS-232 specification is used here rather than the newer RS-449 because most of us are more familiar with the former. In referencing RS-232, the more understandable mnemonics such as DTR, DSR, CTS, and RTS are used rather than the formal CD, CC, CB, and CA terms used in the specification.

NO-MODEM LOCAL DATA CIRCUIT IMPLEMENTATION

Just because some other shop runs local data without modems does not necessarily mean you can. That is an important point to establish now before getting too deeply into implementing no-modem networks. We will be assuming here that your in-house network, front end, terminals, distances, and communication media, topology, protocol, etc., are compatible with a no-modem environment. The only way you may know whether or not your particular application will work without modems is to try it. If you make a test and it is successful, make the test more severe to find out the weaknesses—run longer cable, use smaller-gauge wire, run it near power sources, run it in parallel with other data cables, try several terminal types, etc. Make sure before you commit to no-modem networking as a production service.

SPECIAL DATA CIRCUIT IMPLEMENTATION TOPICS   91

The essence of forming a no-modem network is simply to remove the modem from each end of the circuit and plug the cables back together again. It is really a little more involved than that, but for now we will stay at this level. At the comm area end, once you connect the two cables together, what do you do with them? Probably dump them back under the raised floor to get them out of the way. This is okay and manageable when there are only a few cables to contend with. When a large number of cables are involved, a better method should be used to handle the port cable to circuit cable connection. The best solution is to bring the cables back up out of the floor and use the equivalent of a null modem black box to connect them.

But for a large number of circuits, black boxes are cumbersome, too. So the final solution takes on the form of rack-mounted panels, as shown in Figure 6-1, where the port cables can plug into panel-mounted receptacles. Shown are several racks which provide connection points for several hundred circuits. The two smaller panels at the bottom of each rack are coax patch panels, which will be discussed later. The construction of a no-modem panel can be shown by just one of the individual panels, which can be seen in greater detail in Figure 6-2. The EIA cables from the front end plug into the top panel receptacles, which, in turn, are wired to a socket just below the receptacles. The local network circuits come from the central wiring panel via feeders and terminate on blocks inside the rack as shown in Figure 6-3. (Same philosophy again: everything terminates on the central wiring panel.) This wiring, in turn, leads to the panel, where it is termi-

Fig. 6-1   Rack-mounted circuit panels.

(a)

(b)

Fig. 6-2 Circuit panel detail. (a) Front view; (b) back view.

Fig. 6-3 Local data circuit rack wiring blocks.

nated on the eight-lug terminal strips shown at the bottom of the panel back in Figure 6-2(b). Lastly, the eight-lug strips are wired behind the panel to the same wiring socket as the RS-232 receptacle, as shown in Figure 6-2(b). The link between the EIA wiring and the circuit wiring is then made by inserting a programming plug into the wiring socket to connect the desired signals.

This whole sequence of wiring is rather messy to follow and is summarized in Figure 6-4. In the pictorial and photos, you will notice another wiring socket. There are at least two uses for this socket. One is to plug in transient-suppression devices so that overvoltage transients can be shorted to chassis (protective) ground. You can see an example of one of these further on in Figure 6-19(b), where a zinc oxide varistor is in use. Transient suppression is advisable when a circuit leaves the building.

Another use of this socket is to plug diagnostic equipment into the circuit for troubleshooting as shown in Figure 6-2(a). Here a plug which can be moved to any circuit on the rack is connected to a cable which leads back to the diagnostic center. In this fashion, only one diagnostic port is really necessary to monitor any local no-modem circuit. Studying the diagram and photos should provide enough information to tailor this or a similar setup to your own particular installation.

94 SPECIAL DATA CIRCUIT IMPLEMENTATION TOPICS

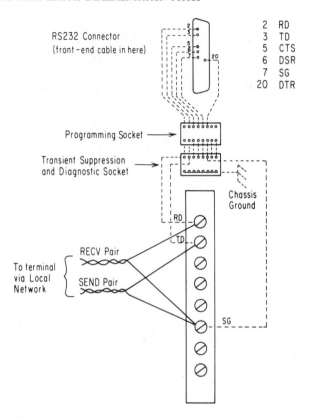

Fig. 6-4 Local data circuit panel wiring.

Now, how do we go about making the thing work? You cannot simply pull out the modem, hook the wires back together, and have the circuit function. The solution is shown in Figure 6-5. The game we will be playing here is to find a way to fake out both the terminal and the front end into thinking they are talking to a modem in the circuit when there really isn't one. This is done differently for different terminals and front ends. Figure 6-5 shows a way to do it for a particular installation, and it may have to be altered for your own.

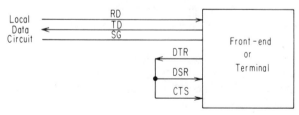

Fig. 6-5 EIA fake-out.

First, both the terminal and front end generally want to see a data set ready (DSR) signal. Both devices produce a signal called data terminal ready (DTR). By shorting DTR to DSR, the condition of the device requiring a DSR signal will be satisfied. Similarly, if clear to send (CTS) is required as a front-end or terminal input, it too can be derived from the DTR signal. At the host end, the signal jumpering takes place by wiring the programming plug appropriately and inserting it into the socket on the no-modem panel. Signal "jumpering" is then repeated at the terminal end by putting jumpers in the terminal connector drop.

The result is that the terminal can talk over a two-pair circuit without modems. There are always special cases in which this scheme will not work and must be modified accordingly. There is a logical place to pull the wool over the terminal's eyes. Don't modify the terminal's connector for the EIA fake-out. Do it in the drop cable from the local network. That way you can maintain standard terminals which will work on either a modem or no-modem circuit without modification.

Local circuits implemented as shown here are very reliable (no modems to fail); they are very easy to troubleshoot because of the connectors and circuit wiring being accessible; and they are easy to track via documentation. It's also quite easy to switch ports when necessary and to group ports or terminal users together on the rack panels. Special circuits can also be easily handled. Most circuits will be simple send-receive wiring. When the need arises to use, say, clear-to-send signaling over the network, it is easily met by adding another pair to the eight-lug terminal strip. Synchronous circuits can be implemented in the same way by running the clock pair out to the terminal.

No-modem sync circuits running over the network described in this book are quite possible. They are a little more tricky, and some help will be provided later in the form of a special adapter assembly to eliminate possible noise problems. Vendors are very skeptical about running their equipment synchronously without modems; they have read me the riot act for doing it. But it works, and I have even thrown in another 1000 ft of wire in some cases to prove that it can work reliably with a very reasonable error rate. In summary, each slot on the circuit panel represents a saving of one pair of modems. That's about $5000 a panel equivalent, and I can assure you that it does not cost that much to build one panel in which only wire and connectors are in use.

COAXIAL LOCAL DATA CIRCUIT IMPLEMENTATION

Chapter 2 briefly discussed the use of a coax patching panel (Figures 2-7 and 2-8) when multiple controllers with their corresponding large number of coaxial cables are in use. In more detail, Figure 6-6 shows the essentials of building patch panels

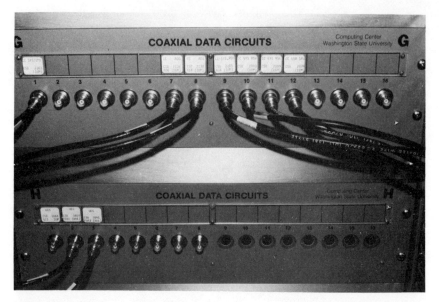

(a)

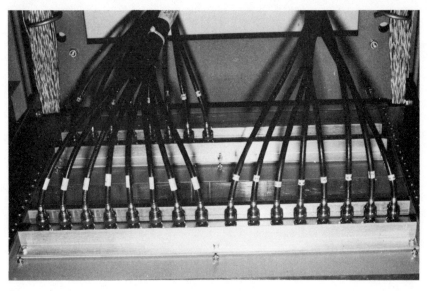

(b)

Fig. 6-6 Coaxial cable patch panel. (a) Front view; (b) back view.

SPECIAL DATA CIRCUIT IMPLEMENTATION TOPICS 97

for coax circuits. The photos are fairly self-explanatory and can be used as a reference for the construction of your own patch panels. Here are some points to consider:

- Notice that a plastic insulator board is used to mount the coax connectors. The material in use here is printed-circuit board (G-10 glass epoxy). This functions to electrically isolate the shield of the coax circuits from the equipment rack, which is a requirement of 327X circuits.
- A narrow Plexiglas plate is mounted to the panel to provide for the attachment of labels which identify the individual circuits.
- When multiple panels are in use, each must be distinguished for the documentation process. The large letters G and H seen in the the photo designate the panel as CDC panel G or panel H. A circuit listing might, for instance, have a circuit number CDC H-01, meaning that the coax is connected to connector 01 on the H CDC panel.
- Coax feeders from the controller connect to the back of the panel as shown in Figure 6-6(*b*). Notice the small cable labels which designate the individual coax lines within the feeder.

If it were not relatively simple to construct coax patch panels, their existence would be doubtful. But since they are inexpensive and simple, they are useful to relieve a little of the frustration normally associated with coaxial circuits. They allow large coax feeders to be installed between the controller and the patch panel rack, thereby eliminating the need to access the controller as frequently and in that way increasing its reliability. Having all the circuit coax readily accessible also allows for rapid installation, moving, and testing of the coaxial data circuits.

The patch panel forms a nice demarcation point between the 327X controller vendor and the network for which the user normally has the responsibility. Vendor finger-pointing at the customer for doing something inside the controller is eliminated. One installation comment may spare you some future frustration: When making the feeder cable between the cable and the patch panel rack, make it long enough that the controller can be moved anywhere within the immediate area. Also, the rack can be located in an out-of-the-way spot from which it is unlikely to be moved. That will save a lot of cable reworking when someone decides to move the controller.

CABLE ASSEMBLIES

A variety of cable assemblies are generally in use in most shops. Unfortunately, the variety also tends to extend to the practices used in building the assemblies. Most of them are built in-house because they are difficult to obtain from vendors. Standards must be followed in building them: use standard cable lengths where applicable; use uniform labeling and documentation; and use good assembly practices. This is an area where unprofessional work crops up, which is somewhat understandable considering the staffing and constraints of some shops.

The information given in Chapter 7 and the Appendix supplements the material given here and should be of some help. Unprofessional cable assembly work from vendors is inexcusable, although you run across it occasionally. Vendors, of all people, should maintain professional standards. If they don't, they should not be patronized. In the following sections, several cable assemblies which should be of interest to most data comm people are described. The examples illustrated should provide enough information for readers to either copy the assembly shown or adapt it to their own particular needs. The RS-232 conventions used here can easily be converted to RS-449 if needed.

Most of the illustrations in this section are electrical schematics for the particular cable assembly under consideration. How do you go about building the cables shown here? Figure 6-7 provides some useful information, which, when added to the details discussed in Chapter 7 and the Appendix, should get you into the cable assembly business. All the assemblies shown here can be made by using crimp-type connector hardware manufactured by AMP Incorporated (AMP), and they are noted by part number in the schematics. Use Figures 6-8 and 6-9 as examples of how the individual wires are drawn. Figure 6-8 shows a cable made from two *twisted pairs;* this is my own symbol for twisted pair to allow information such as wire color to be added unambiguously. Figure 6-9 shows *nontwisted pair* cable in use.

Four-Wire Extender Cable

Terminals running over a two-pair no-modem network may require an extension cable (Figure 6-8). The cable simply extends the two-pair circuit as desired and provides the appropriate jumper wiring in the terminal receptacle to make the terminal think it is connected to a modem. Extension cables should be built in standard multiples such as 50 ft (15 m) and 100 ft (30 m).

Ten-Wire Extender Cable

The ten-wire extender (Figure 6-9) fulfills the same function as the four-wire extender except that it is used for extending the terminal further from a *modem*. Most of the basic signals are extended, which allows the cable to work in most applications.

No-Modem Terminal Drop—Standard Async

On a local no-modem network several terminal types, each requiring its own special drop cable, may be in use. Three common drop cables are illustrated in

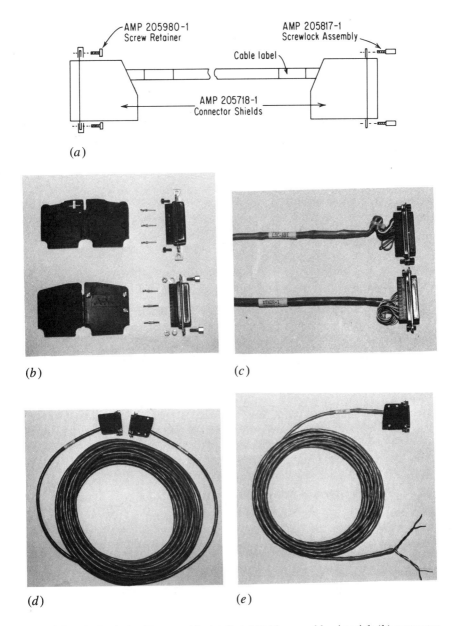

Fig. 6-7 Mechanical cable assembly detail. (*a*) Cable assembly pictorial; (*b*) connector components; (*c*) assembly without shield; (*d*) extender cable; (*e*) terminal drop cable.

100 SPECIAL DATA CIRCUIT IMPLEMENTATION TOPICS

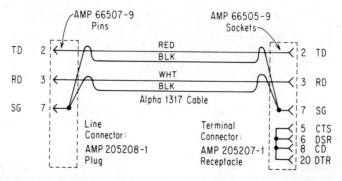

Fig. 6-8 Four-wire extender cable.

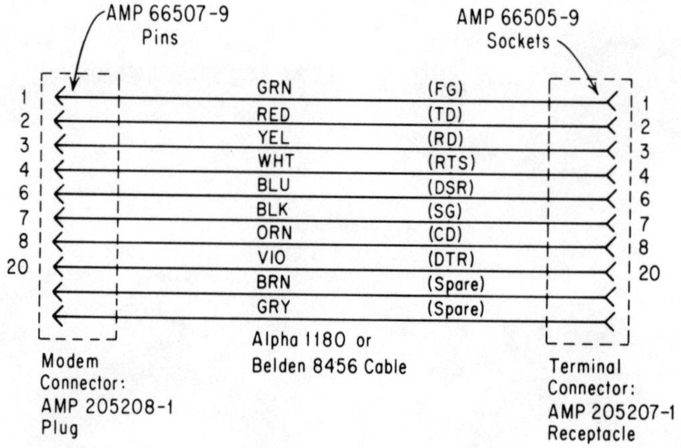

Fig. 6-9 Ten-wire extender cable.

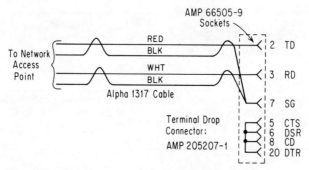

Fig. 6-10 No-modem terminal drop—standard async.

this section, beginning with Figure 6-10. Here a drop cable for standard asynchronous-type terminals is shown. The drop cable brings the two data pairs from the network to the terminal and also provides the jumper wiring needed for the no-modem application.

No-Modem Terminal Drop—RTS Async

Some terminals are buffered for data transmission and may require RTS-CTS signaling to throttle the data from the host. To accomplish this, a third pair is added to the standard async drop. RTS from the terminal is wired back over the local network and eventually is wired to the front-end CTS input via the no-modem wiring panel. See Figure 6-11.

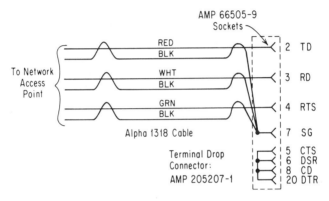

Fig. 6-11 No-modem terminal drop—RTS async.

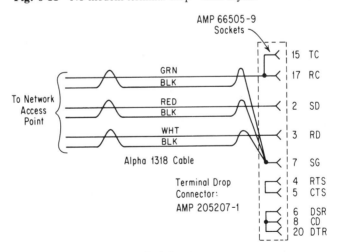

Fig. 6-12 No-modem terminal drop—sync.

No-Modem Terminal Drop—Sync

Synchronous terminals can be run in a no-modem environment by adding the clock signal to the basic terminal drop cable (Figure 6-12). Depending upon several factors such as terminal type, data rate, and line length, a special adapter may have to be added to this terminal drop to improve the circuit performance. The adapter is described in the next section.

MUX to 103 (113) Cable

Just in case you have a need to provide dial access to something like a FDM or statistical multiplexer circuit, the cable shown in Figure 6-13 should either do the trick or get you confused if your application won't allow this particular configuration. The cable is inserted between the 103 or 113 data set and the multiplexer.

Long-Haul to Short-Haul Cable

If the last cable described wasn't confusing enough, then the one shown in Figure 6-14 may get you climbing the wall. Under certain circumstances you may have need to connect short-haul modems to a long-haul modem to form a remote short-haul network. To do this requires carrying the clocking through to the short-haul modem.

ADAPTER ASSEMBLIES

Adapter assemblies, like cable assemblies, are found in many data circuit implementations serving a variety of purposes. The format presented here is the same

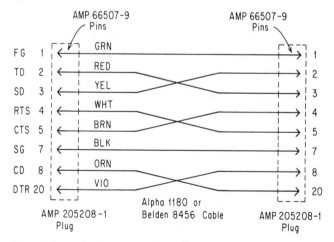

Fig. 6-13 MUX to 103 (113) cable.

SPECIAL DATA CIRCUIT IMPLEMENTATION TOPICS   **103**

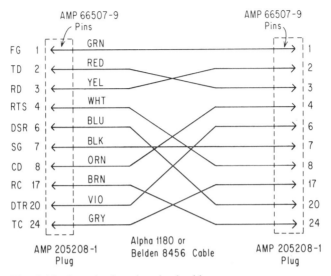

Fig. 6-14 Long-haul to short-haul cable.

as for the cable assemblies: general mechanical information applicable to all adapters, followed by electrical schematics for the particular function desired. Figure 6-15 gives the mechanical information for construction of the adapters along with photos to better illustrate what is involved. The adapters illustrated are constructed with solder-type RS-232 connectors to form a rigid assembly, rather than by using the crimp-type components. Not all the pin connections shown in the schematics are electrically necessary; they are used for mechanical strength of the adapter assembly. Those which are required for circuit operation are denoted with their appropriate EIA mnemonic. The plastic dust cover shown in the photos is simply a piece of Plexiglas bent and formed with a hot-air gun.

FDM Adapter

The FDM adapter (Figure 6-16) simply allows the host or front end to know when carrier is lost on some particular FDM channel. Carrier detect dropping drops CTS and DSR to signal that the circuit is no longer in operation.

Carrier-Controlled CTS Adapter

Analogously to the RTS terminal drop cable assembly, you may have need to throttle data over a modem link via carrier control. This adapter (Figure 6-17) causes a loss of carrier to drop clear-to-send to the front end, thereby stopping host output. At the terminal end, the RTS from the terminal needs to be set up to toggle carrier.

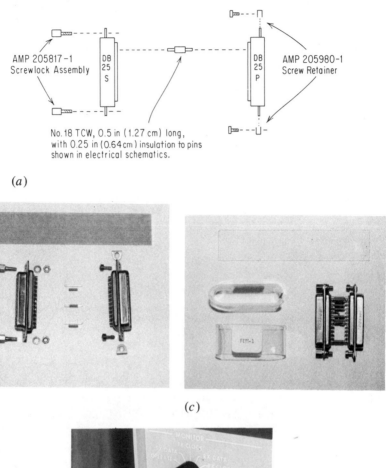

Fig. 6-15 Mechanical adapter assembly detail. (*a*) Adapter pictorial; (*b*) adapter components; (*c*) dust cover; (*d*) installed adapter.

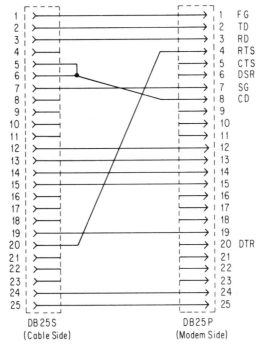

Fig. 6-16 FDM adapter.

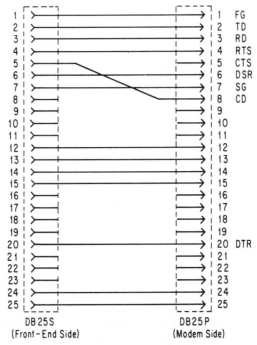

Fig. 6-17 Carrier-controlled CTS adapter.

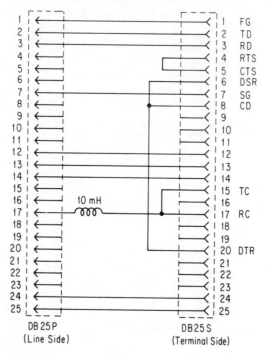

Fig. 6-18 Sync noise reducer.

Sync Noise Reducer

After you implement the synchronous terminal drop shown in Figure 6-12, you may have an unacceptably high link error rate. There are several ways to go about improving the performance, but in my experience, the simple addition of an inductor (Figure 6-18) works about the best. A small 10-mH signal inductor put in series with the clock line does wonders for most synchronous drops running without modems. The adapter will work on data rates up to 10 Kbaud without any changes. I have run sync drops over 1000 ft (300 m) using this adapter with reasonable error performance. If your circuit cannot be improved with this device, there may be something more fundamentally wrong.

MISCELLANEOUS CIRCUITS

Overvoltage Suppression

Extending the local network outside the building forces one to consider overvoltage suppression (Figure 6-19). Inside the building, either lightning or static is not much of a problem for the relatively low impedance local data circuits. Special conditions may cause internal problems, but overall, there is not much to

SPECIAL DATA CIRCUIT IMPLEMENTATION TOPICS 107

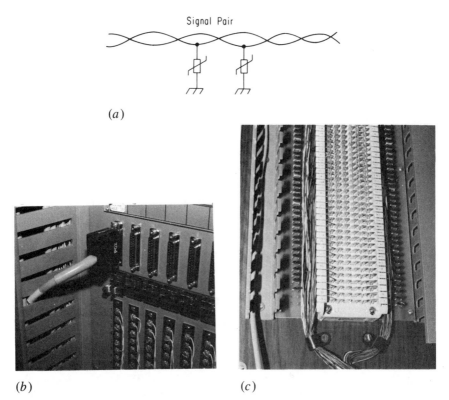

Fig. 6-19 Overvoltage suppression. (*a*) Circuit wiring; (*b*) VDR mounted on IC header; (*c*) block suppression detail.

worry about inside the structure. Interbuilding transients are a different matter. Lightning strikes can set up considerable voltage differences between facilities and can do damage to unprotected circuits. We are talking here only about your own local data circuits, not Telco's. Telco always installs transient protection at the building entrance point to protect data circuits. You must do the same thing.

Of the several devices available to do the transient protection job, we will consider only one. For reasons of price, utility, and simplicity, the zinc oxide varistors shown in Figure 6-19 will perform well. The device is a nonlinear resistor which varies its resistance inversely as the applied voltage changes. It is known as a voltage-dependent resistor (VDR), and it is available from a variety of electronic supply houses. For data applications, you will need to use low-voltage VDRs; around 50 V is a useful value. You probably could get by with putting one VDR between the individual wires of a data pair. That would keep the differential voltage limited with respect to only the wires, not the wires and some other reference. As a matter of practice, I use one VDR per wire, i.e., two per data pair, and reference them to a good earth or green-wire ground. That keeps everything relative to the building ground.

108 SPECIAL DATA CIRCUIT IMPLEMENTATION TOPICS

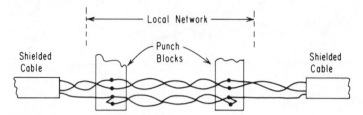

Fig. 6-20 Shielded twisted pair circuit.

Shielded Twisted Pair Circuits

Many data terminals have rather long transmission lines where shielded twisted pair is in use (Figure 6-20). The reasons for the use of shielded twisted pair are valid because the data cable may pass through some rather harsh conditions. The problem here is how you connect this type of transmission line to your local network, whose punch blocks are not well acquainted with this media. Actually, your first problem will be with the vendor, who will probably tell you what a horrible thing it would be to unshield the data pairs when you connect to your local network. After you demonstrate that it works and put in another 1000 ft (300 m) of wire to show your confidence, the vendor rep will settle down.

I am assuming here that your network is clean and well designed as suggested in this book. Your application may be something like hooking up a remote point-of-sale terminal to a processor in your facility. To do this, connect the shielded twisted pair to the network access point as shown in Figure 6-20. The shield or drain wire of the cable consumes one full pair, as shown, to carry the shield conductor across the network. Once the cable reaches its destination, the reverse procedure is used. Continuitywise, the circuit is the same as if the shielded cable were used throughout. Noisewise, the data pair within the network is exposed because of the lack of a shield. But since your network is clean, little noise is coupled into the signal, and chances are very good that no one can tell the difference in performance.

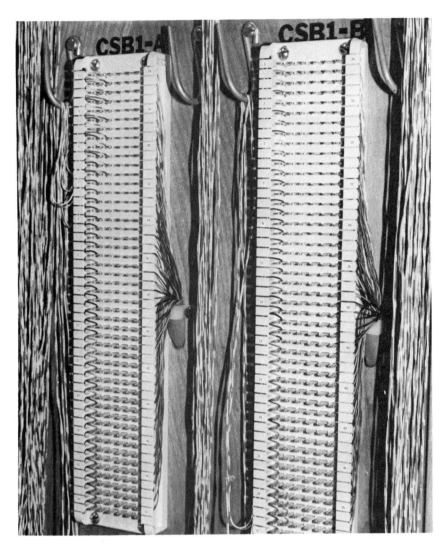

CONVENTIONS AND PRACTICES

CHAPTER 7

110   CONVENTIONS AND PRACTICES

Tired of constructing buildings and data circuits? We're done! Now we need to establish a few conventions and practices which you should find useful to do the job more professionally. Many conventions have already been mentioned, but here we want to concentrate on them and dig more deeply into the reasoning behind them. Why have conventions anyway? Here are a few reasons:

- There are typically several people in the data comm shop. You can't have each of them doing his or her own thing and expect any consistency. It's counterproductive to allow people to set their own conventions.
- How about training new personnel? It's much easier if you have documented policies, procedures, and also conventions. New people should have a reference book to lean on to help them over the hurdles; it's tough enough to get started without causing problems by setting new conventions contrary to established methods.
- Consistency has been mentioned, but it bears repeating. You really do need to maintain consistency or you will be wasting considerable time and cause frustration for those who work with the network.
- Conventions ensure that the job is being done correctly. It is often not obvious, especially to new people who lack experience, why certain things are done the way they are. If people have to follow a convention, at least they can raise the question why they must follow it. Without one, they will set their own courses and perhaps do the job incorrectly, which, in turn, may cause problems later on.
- Conventions help overcome the inexperience just mentioned. Newcomers progress at a much faster rate by using established conventions rather than learning, the hard way, by repeating mistakes over and over again. Why repeat mistakes someone else has already committed?

COLOR CODING

Telco Color Coding

The telephone cable used for all twisted pair networking is color-coded into five major and five minor colors for pair and group identification as shown in Table 7-1. The colors are prioritized. This scheme provides for the unique coding of all 50 wires in a 25-pair group. You will notice that each individual wire has two colors and that one of the two predominates. The cable is coded in groups of five by using the major and minor colors. The first group is the white group. White/blue coding is for pair 1, white/orange coding for pair 2, white/green for pair 3, etc. The fifth group is violet/blue for pair 21, violet/orange for pair 22, violet/green for pair 23, etc. Notice that we are speaking of wire-pairs here, not individual wires.

TABLE 7-1 Telco Color Coding

Priority	Major colors	Minor colors
1	White	Blue
2	Red	Orange
3	Black	Green
4	Yellow	Brown
5	Violet	Gray

The individual wires within each pair have their colors reversed to provide identification down to the wire level. The color coding for pair 1, for instance, is white/blue for one of the wires, and blue/white for the other. The first color given is the most predominant color of the two. Table 7-2 shows all the color coding down to the wire level for 25 pairs.

By the scheme described above, the wires in 25-pair cable are uniquely identified. What about 50-pair or larger? If you strip the jacket off a large cable, you will notice that each group of 25 pairs has a color-coded thread wrapped around it. The same color coding is used for the thread to identify the groups of 25 pairs. In a 200-pair, for example, a white/blue thread would be the first group of 25, white/orange the second, etc. When in doubt, just follow the major and minor color sequence to determine priority.

Several variations are used by cable manufacturers to identify the 25-pair grouping. Some use colored thread, and others may use colored tape. Just recognizing that the identification is by color should help you to figure out the scheme used by the manufacturer.

Color: Functional Association

For cable assemblies and other wiring jobs it is useful to associate colors with certain functions. In other fields, some functional associations which are useful for data communications are in use. People in electronics use the resistor color code, which identifies the numbers 0 to 9 with a unique color as shown in Table 7-3. If you must build a special cable with numerical coding, you will find it useful to follow the resistor color code. If you use the code as your in-house standard, all your numerically coded cable will be consistent.

Other less standardized color codes which have come into use in the trades can be adapted to your shop practices. Some of the more common ones are listed in Table 7-4.

TABLE 7-2 25-Pair Color Coding

Pair no.	Color code	Pair no.	Color code
1	White/blue Blue/white	14	Black/brown Brown/black
2	White/orange Orange/white	15	Black/gray Gray/black
3	White/green Green/white	16	Yellow/blue Blue/yellow
4	White/brown Brown/white	17	Yellow/orange Orange/yellow
5	White/gray Gray/white	18	Yellow/green Green/yellow
6	Red/blue Blue/red	19	Yellow/brown Brown/yellow
7	Red/orange Orange/red	20	Yellow/gray Gray/yellow
8	Red/green Green/red	21	Violet/blue Blue/violet
9	Red/brown Brown/red	22	Violet/orange Orange/violet
10	Red/gray Gray/red	23	Violet/green Green/violet
11	Black/blue Blue/black	24	Violet/brown Brown/violet
12	Black/orange Orange/black	25	Violet/gray Gray/violet
13	Black/green Green/black		

Color Mnemonics

In documentation, you soon get tired writing down all the color coding in longhand. It is useful to use three-letter mnemonics to abbreviate the colors as listed in Table 7-5.

TABLE 7-3 Resistor Color Code

Number	Color	Number	Color
0	Black	5	Green
1	Brown	6	Blue
2	Red	7	Violet
3	Orange	8	Gray
4	Yellow	9	White

TABLE 7-4 Color-Function Association

Color	Function
Black	Signal ground
Red	Positive supply voltage, commonly +5 V dc
Green	Chassis or earth ground
Red/green	Transmit pair
Yellow/black	Receive pair

TABLE 7-5 Color Mnemonics

Color	Mnemonic	Color	Mnemonic
Black	BLK	Green	GRN
Brown	BRN	Blue	BLU
Red	RED	Violet	VIO
Orange	ORN	Gray	GRY
Yellow	YEL	White	WHT

A couple points need clarification. You will find that the telephone company frequently uses slate rather than gray for that color name. I find gray preferable because GRY has more meaning for most people than SLT would have. Another point is that people sometimes get sloppy and use GRN for ground when they should use GND.

EIA Color Code

In working with EIA signals in data communications, I have found it useful to associate certain EIA signals with colors just to try to maintain a little consisconsistency. Table 7-6 lists the more common EIA signals with suggested colors. I am

114 CONVENTIONS AND PRACTICES

not aware of any formal EIA coding in existence. You too may find this a useful addition to your practices. You will also find that it is sometimes impossible to follow this coding religiously.

TABLE 7-6 EIA Color Code

Pin	Signal	Color	Function
1	FG	GRN	Protective or frame ground
2	TD	RED	Transmit or send data
3	RD	YEL	Receive data
4	RTS	WHT	Request to send
5	CTS	BRN	Clear to send
6	DSR	BLU	Data set ready
7	SG	BLK	Signal ground
8	CD	ORN	Carrier detect/carrier on
20	DTR	VIO	Data terminal ready

EIA MNEMONICS

This section lists some of the EIA signal mnemonics which are in common use. In practice, most people don't use the formal AA, BA, BB, CA, etc., mnemonics used in the RS-232 specification because they are content-meaningless. The intent of this section is to list the more common abbreviations in use so that some consistency is maintained in this area. The newer RS-449 specification did a much better job in regard to signal mnemonics. It will be difficult for a lot of us to get into the habit of using the newer names, however, because the old ones are so imbedded in our way of referencing the signals. We should all try to adopt the RS-449 names in the future. I thought about doing that in this book but did not because of the confusion it might cause. Table 7-7 lists some of the more common RS-232 signals and their unofficial names with a reference to the newer RS-449 conventions.

PUNCH BLOCK PANEL WIRING CONVENTIONS*

Figures 7-1(*a*) and 7-1(*b*) provide useful wire routing conventions for the wiring panels described in this book. Some of the conventions shown are the same as those used in telephone company standard wiring practices, but others differ considerably. Close attention to these figures should eliminate confusion and ensure that neat and consistent wiring is maintained. Here are some suggestions:

---

* This section, and the following three sections, are excerpted from *Application Notes 2, 3, 4,* and *5,* respectively, copyright © 1981 by Computer Energy, Inc.

TABLE 7-7 EIA Mnemonics

Formal RS-232	Informal RS-232	Formal RS-449	Definition
AA	FG	n/a	Frame (or chassis) ground
BA	TD,SD	SD	Transmit or send data
BB	RD	RD	Receive data
CA	RTS	RS	Request to send
CB	CTS	CS	Clear to send
CC	DSR	DM	Data set ready Data mode
AB	SG	SG	Signal ground
		SC	Send common
		RC	Receive common
CF	CD,DCD	RR	Data carrier detect Receiver ready
DB	TC	ST	Transmitter clock Send timing
DD	RC	RT	Receiver clock Receive timing
CD	DTR	TR	Data terminal ready Terminal ready
CG	SQ	SQ	Signal quality
CE	RI	IC	Ring indicator Incoming call

1. When both telephone company and customer in-house wiring are done on the same panel system, use different wire colors to denote ownership, e.g., orange/white for Telco, blue/white for customer.
2. Keep wire taut for neat appearance and to reduce the chance of intertwining wire-pairs.
3. It is normal for wire to cross in the wire guide area.
4. Never wire directly between blocks, no matter how close the blocks are to one another. Always use the wire guide and loop.

PUNCH BLOCK NAMING & LABELING CONVENTIONS

This section discusses some of the considerations which should be given to the naming of punch blocks used in communication networking. Most often, this aspect of hardware labeling is given little attention even though it eventually becomes intimately linked with the documentation describing the network signal flow. As the network grows, it becomes more apparent that an efficient naming convention can appreciably enhance network documentation. The conventions

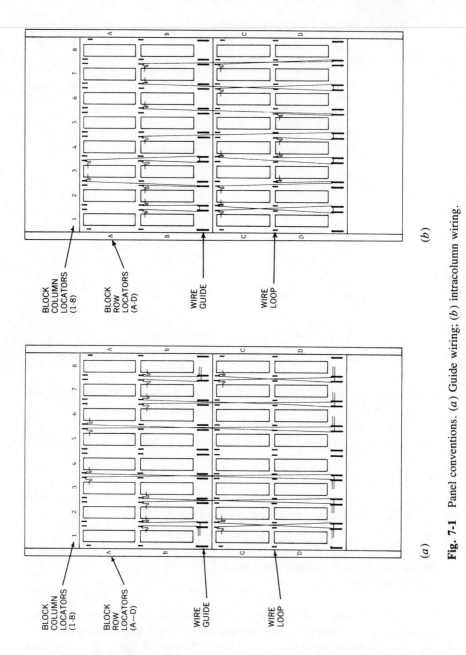

Fig. 7-1 Panel conventions. (*a*) Guide wiring; (*b*) intracolumn wiring.

described here will be useful in selecting names for punch blocks on the central wiring panel or blocks located at the terminal end.

Name Characteristics

The criteria for a good naming convention would include the following:

1. The name mnemonic should be short because there is generally a limited amount of labeling space on the block mounting panels. In addition, suffixes are desirable for the name, which further restricts the name length. Short names also permit larger block lettering for greater ease in working with the network visually. It turns out that four or fewer characters are a good length for a mnemonic.
2. The convention must take into account both how individual signals (e.g., SEND, RECV) flow and how groups of these signals flow. Network documentation may largely describe the circuits for individual terminals, but there may be a common group (such as a cluster of terminals in a room) of signals needing description. Because of this, the convention needs to have a two-dimensional characteristic, requiring suffixes.
3. The convention should be adaptable to documenting the network via a computer file. Networking can be quite dynamic, and frequent changes may be required. Performing the documentation via a computer file is an important feature for maintaining neat and accurate records.

Selecting the Name

The name mnemonic should have from one to four alphanumeric characters. Common characters should be used to eliminate problems caused by printing special ones.

Name Suffixing

Suffixes must be added to name mnemonics to allow easy identification of individual signals and groups of signals. The simple addition of an alphabetic character suffix to a particular name mnemonic allows several blocks serving a common usage to be grouped. For instance, if two blocks were required to wire a room full of terminals, one might select TERM as the name mnemonic and then suffix the two block labels as TERM-A and TERM-B. This then defines the group of blocks providing service to the room. Now, any particular signal might go through several blocks in series to implement the overall data circuit. To document this, a number must be added to the first suffix to trace the signal flow.

Continuing the example, suppose the terminals in the room use two blocks in series for the overall circuit as shown in Figure 7-2. Adding number suffixes to

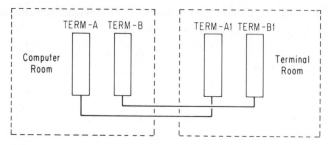

Fig. 7-2 Block naming example.

one set of blocks (A1, B1) uniquely identifies the blocks and defines the signal flow. Generally, one would suffix numbers increasing in value away from the source of service.

Generalized Block Naming Conventions

In summary of the above comments, Table 7-8 defines the naming conventions.

TABLE 7-8 General Block Naming Conventions

		Signal flow ⟶		
	NAME-A	NAME-A1	. . .	NAME-An
	NAME-B	NAME-B1	. . .	
Group	NAME-C	. . .		
flow	.			
	.			
↓	.			
	NAME-a	NAME-a1	. . .	NAME-an

Guidelines

Here are several suggestions for implementing the conventions:

1. When labeling the block, starting the name at the left will leave room for future suffixing without reworking the original name, as shown in Figure 7-3.
2. For circuit documentation, the individual block wire-pair information is suffixed to the name mnemonic. For example, a wire-pair connected to block pair position 10 on block TERM-B1 would be documented TERM-B1-10.
3. The width of a standard 66B-type punch block will make 0.5-in- (1.27-cm-) high lettering a usable standard. Commonly available plastic peel-off lettering works well for this purpose. This size permits seven positions for labeling the block. Some possible name styles are:

Fig. 7-3 Block label example.

NAME-A	NAME-A1	NAME-A2	...
EX1-A	EX1-A1	EX1-A2	...
EX2-A	EX2-A1	EX2-A2	...
EX3-AA1	EX3-AA2	EX3-AA3	...

4. It is more consistent to add a letter suffix to every block even if no signal groups exist. For example, XMPL-A instead of XMPL will be more consis-consistent. This also makes it easy to add more blocks in the future to form groups of blocks without altering the original labeling and documentation.

PUNCH BLOCK TAB NUMBERING CONVENTIONS

Numbering the tabs on punch blocks is necessary to locate and document the wire attached to the block. It is not necessary to number all tabs. Doing so is probably the best way, but the main reason for not numbering all tabs is that it is time-time-consuming. Too few labels, however, make using the blocks difficult and increase the chance of making a wiring or documentation error. A compromise is needed to reduce the labeling time while maintaining a workable end result.

Most data circuit wiring consists of two data pairs: send pair and receive pair. Here this physical pairing is used to adopt the compromise tab numbering convention. Block installation is customarily via 50-pair blocks mounted vertically, with the top left becoming pair 1 and the bottom right pair 50. Pair wiring is also customarily done RECV1, SEND1, RECV2, SEND2, etc., starting at pair position 1 on the block. SEND/RECV is from the user's viewpoint. Also, for 2-pair circuits, one pair position is unused on each side of 50-pair blocks. With these comments in mind, tab numbering can be adopted as shown in Figure 7-4. Note that each number is associated only with a RECV pair. This tends to form 2-pair circuit groups over the block, each easily distinguishable. This scheme can also be used for other than 50-pair blocks and blocks used horizontally.

SEND/RECV CONVENTIONS

Data circuits using the common SEND and RECV data pairs are neatly implemented via punch-block-type wiring components and systems. When these components are used, it is easy to say arbitrarily that the RECV pair will always be on top (vertical blocks) followed by the SEND pair for that circuit. This is indeed

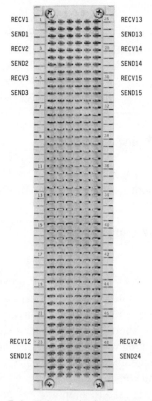

Fig. 7-4 Tab numbering conventions.

a good convention to follow to maintain consistency and ease of working with the network. However, ambiguity becomes apparent when you ask: "Whose RECV or SEND pair is it? Is it ours? Is it the telephone company's?" After deciding that, you must then ask how far across the network the viewpoint of ownership holds. At the central wiring panel, Telco circuits terminate along with all building circuits and adjacent building circuits. Here the problem is obvious: Is a RECV pair on the central wiring panel a receive signal from Telco's viewpoint? From a local terminal's viewpoint? An adjacent building terminal's viewpoint? The front end's? To resolve this dilemma requires some rigid definitions:

Network

The twisted pair network under consideration is defined as including the central wiring panel, all associated terminal wiring blocks, and all connecting cable between them. The wire from the terminal to the terminal block is not part of the network. The terminal block is viewed as the entry point to the network for that terminal. A Telco circuit is not part of the network. The first block it attaches to would be Telco's entry point into the network. This scheme would also apply to all other entities such as front ends and modems. The network may have two or more central wiring panels in separate buildings, all linked together. These may or may not all be considered part of the same network. This may depend upon ownership, politics, and type of systems connected to the network.

Server

Generally, the central wiring panel system would be installed for the purpose of connecting various terminals and other devices to some central data processing facility which provides services to all or most attached network devices. From the viewpoint of the central DP facility, it is a "server" of DP services to various attached devices. Most likely, the server is also the owner of the attached network.

Using these definitions for network and server, a definition for the SEND and RECV pairs can now be made:

SEND and RECV signals are from the server's viewpoint anywhere within the network.

For the majority of applications, this definition works well, but there are a few instances when an alternate definition may be better. Care must be exercised to use a definition appropriate to most uses of the network.

Sometimes color-coded four-wire cable is used to connect equipment to the network. The general rule is to connect as follows:

SEND	Red/green
RECV	Yellow/black

Here the definitions for the SEND and RECV coding would be from the viewpoint of the equipment being connected to the network. An example is provided in Figure 7-5 to help clarify these definitions.

FLOOR TILE CUTTING STANDARDS

Do you standardize the way you cut your floor tiles? If you don't, it will not be long before you have a lot of unusable tiles. Figure 7-6 shows several basic tile

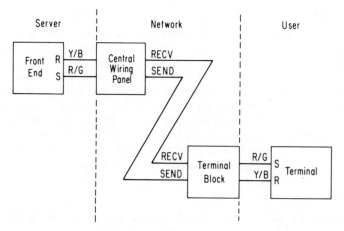

Fig. 7-5 SEND/RECV definition.

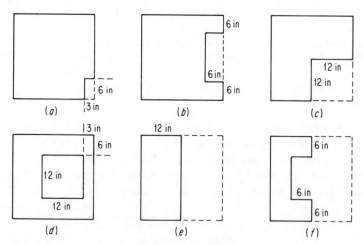

Fig. 7-6 Basic floor tile cuts (2-ft tiles).

CONVENTIONS AND PRACTICES   **123**

cuts. By using them as much as possible, you will avoid a lot of tile butchering. Note the following in regard to the basic tile cuts:

- *Tile a*
  This tile is useful for bringing one or a few cables up to a terminal or device. The hole size and shape allows for the larger "bus and tag" type connectors to pass through the hole.
- *Tile b*
  Two adjacent *b* tiles will open a one-square-foot hole across a floor support rail.
- *Tile d*
  This tile is cut off-center so that rotating it moves the relationship between the hole and equipment over it.
- *Tile f*
  Note that tile *f* combined with *e* forms a different version of tile *d*.

Many people lay out the equipment without regard to floor tile cutouts and then, after the layout is done, custom-cut the tiles for that layout. I guess that's all right if you have an adequate floor tile budget, but it is wasteful. You never can get away from some custom-cutting. A good compromise is to do the equipment layout with the basic tile cuts in mind and supplement them with a few custom cuts as required. You may have to alter the basic cuts for your particular tiles.

MISCELLANEOUS PRACTICES

1. On some systems where there might not be very elaborate diagnostic facilities, you can get a feel for your data circuit performance by using your system information on port data transmission. In block or packet communication, for instance, if you know the number of units transmitted and the number of units retransmitted because of errors, you can derive a quality number as follows:

$$Q = \frac{\text{number of units transmitted correctly}}{\text{total number of units transmitted}}$$

This will provide you more information than old programs like EREP, which just gives you the errors without any feel for the overall line utilization.

2. When deciding on a mnemonic for labels and documentation, choose one to minimize future changing. For instance, if you select DEC or IBM as a prefix for circuit numbers because the cabling connects to that particular vendor's equipment, you will have to change all your labels and documentation if you switch vendors. LDC (local data circuit), in this case, would be an example of a general prefix to avoid the problem.

3. If you build the LDC or CDC racks to control the cabling for these circuits more efficiently, your first problem will be the philosophy to adopt for assigning space on the racks. With all those panels, it is quite tempting to assign panels to various end users or use some other orderly grouping. In practice, that doesn't work. When rack space becomes a premium or users move around and disrupt your order, you will soon discover that ordered panels are too difficult to maintain. This is a case where chaos is better than order. The best utilization strategy seems to be a randomly assigned panel space.

DOCUMENTATION
CHAPTER 8

Paperwork! Now we are getting into the stuff we all hate to do but can't live without. I don't know of any way to paint a rosy picture of this subject, it's simply one of the overhead tasks that comes with the job. It is interesting to observe the differing managerial philosophies on documentation; they range from almost none to so much that the staff is literally buried in it. When documentation is not required, sooner or later the painful discovery will be made that a shop can't be run without some established documentation procedures. This is especially true when you are in the business of providing services to customers. Without documentation, the staff seems uncoordinated and flounders around trying to solve problems and provide support. This doesn't look good to the customers. There are always some customers who do not understand your normal problems working in data comm because of their own inexperience; you do not need to add to the problem by ignoring procedures which both you and your customers can see would be of benefit.

Documentation is one of those areas in which performance can be judged by less technical people. When a technical problem arises, you can undoubtedly do a snow job on some customers who will judge your performance by how fast you get them back into operational status. They won't judge you on how you went about doing it; they cannot as easily judge that aspect. Documentation, however, is another matter; your customers can certainly see the reason for your inability to trace a circuit, find systems-related information, or find technical information needed to solve their problems. On the other hand, customers can also see a documentation problem when they have to wade through a bureaucracy to get a problem solved. "Sorry, sir, I cannot look into your problem until the proper paperwork is completed!" That should not happen either directly or indirectly. The customers couldn't care less about your internal procedures; they just want their problems solved.

You need a compromise to these extremes of handling the paperwork. Each shop needs to stand back and look at its documentation practices to find a way to implement procedures that add to the efficiency of the operation without creating a paper monster. The aim of this chapter is not to show how to implement proper documentation procedures for your shop; you must decide that. You know your particular situation and the politics involved, and you are in the best position to appraise the documentation needs. About all I can do here is list a few things from my own experience which you may want to include in your documentation library and some other items related to the operational and network paperwork. What you need to do is combine this information with your existing documentation and make a decision on what is appropriate for your own shop and its particular constraints and operation.

The tough part is deciding what documentation is really good to have versus being a burden on the operation and staff. My parting advice on the philosophy of documentation is to the newcomers (the old-timers already know this). You really do need to practice good documentation procedures. The sooner you realize this, the more efficiently your operation will run. But just don't get too carried away with it!

THE DATA COMMUNICATIONS REFERENCE LIBRARY

You need a library of sorts. It may be shelves full of material, filing cabinets, or both depending upon the size and scope of your operation. It should be located near the data comm activity where it is needed most, and it should be organized for ease of locating information. About the only information that may not be located near your network activity would be the archive material. The documentation classifications that seem appropriate for a data comm library are administrative, technical/reference, data comm literature, and archive material. These subjects are covered in this section. In later sections you will find material which doesn't necessarily fit into a library classification: the operational and network documentation for your shop. This type of documentation, such as forms and circuit listings, may be somewhere other than in the library.

Administrative Library Material

The administrative section of the library should contain items describing the interaction of the network group with other groups, how the network group generally goes about its business, what agreements the network staff must operate under, and similar items. We are talking here of policies, procedures, and agreements which are probably established by higher management or at least with its knowledge and approval. These materials need to be in the data comm library so they are readily available to the network staff. Probably not included in this classification would be such items as the procedures to fix a certain type of modem. Those procedures are generally written by the network staff or manager with only the manager's knowledge and approval; higher management doesn't really care about that level of detail. Such items belong more in the operational and networking documentation described in a later section. Now that we have a taste of the type of things appropriate for the administrative section, the following goes into the subject in a little more detail.

1. *Policies and Procedures*

   The policies and procedures may be divided into two classifications:
   - *Divisional policies and procedures* may cross group boundaries and responsibilities. Policies are needed to define areas of divisional versus network responsibilities so that it is clear who is responsible for what. These policies would no doubt be established by divisional management. Another area under this classification would be established policies which relate to

the handling of the customer base. Management must establish these policies so that the support staff knows how it is supposed to respond to situations normally arising in its dealings with the customers.
- *Network policies and procedures* are pertinent to only the network group. They are of a general nature, and they establish how the network group functions in its activities. They are at a level that upper management understands and supports, which lets the network staff know how upper management views its function. These policies and procedures are probably written by the network management and established as policy with upper management's approval. The following are some examples of these policies and procedures:
  - Defining general areas of responsibility for the network group
  - General equipment maintenance policies
  - Network troubleshooting policies
  - Policy on customers hanging nonstandard devices on the network which you may not support
  - What to do with customer-owned equipment located in your shop

2. *Agreements*

   Again, there are two obvious categories of the various agreements generally in use in the network area: vendor agreements and customer agreements. You should include them in your library for staff reference. This is an area where management personnel are more knowledgeable on the subject because they are the ones who normally deal with these matters. When the network staff comes onto a problem affected by some agreement, it is nice to have the material readily available for reference. Generally, the network staff is not well acquainted with this type of material (and shouldn't necessarily have to be), so having it close by avoids a potential problem area.

Technical/Reference Library Material

1. For knowing telephone company circuits and practices, a set of Bell Technical References is a good place to start. These references, which are published by AT&T, are a very useful addition to your library. An address is given in the Appendix; see AT&T. You can start with the Bell catalog, *PUB 40000,* which lists all the publications. You don't need all of them, but you should get those pertinent to circuits and equipment you normally use in your operation. There is an individual publication for each type of Bell modem, such as 103, 202, and 208. The publications are very useful when you need to order equipment from either Telco or other vendors. The references are the standard for a particular data set with which second sources to Bell comply. The following are other references that deserve your attention:

- PUB 41004: *Data Communications Using Voiceband Private Line Channels*
- PUB 41005: *Data Communications Using the Switched Telecommunications Network*
- PUB 41008: *Transmission Parameters Affecting Voiceband Data Transmission*
- PUB 41021: *Digital Data System Channel Interface Specifications*
- PUB 43201: *Private Line Interconnection*
- PUB 43401: *Transmission Specifications for Private Line Metallic Circuits*
- PUB 53401: *Color Combinations and Uses, Wiring and Cabling General Equipment Requirements*

2. Another useful set of reference materials is the group of EIA specifications related to data communications. An address from which you can acquire this material is given in the Appendix. Some publications you might consider are the following:

- Industrial Electronics Bulletin No. 5: *Tutorial Paper on Signal Quality at Digital Interface*
- Industrial Electronics Bulletin No. 9: *Application Notes for EIA Standard RS-232-C*
- Industrial Electronics Bulletin No. 11: *Fault Isolation Methods for Data Communications Systems*
- Industrial Electronics Bulletin No. 12: *Application Notes on Interconnection Between Interface Circuits Using RS-449 and RS-232-C*
- EIA Standard RS-232-C: *Interface Between Data Terminal Equipment and Data Communications Equipment Employing Serial Binary Data Interchange*
- EIA Standard RS-269-B: *Synchronous Signaling Rates for Data Transmission*
- EIA Standard RS-334: *Signal Quality at Interface Between Data Processing Terminal Equipment and Synchronous Data Communications Equipment for Serial Data Transmission*
- EIA Standard RS-363: *Standard for Specifying Signal Quality for Transmitting and Receiving Data Processing Terminal Equipments Using Serial Data Transmission at the Interface With Non-synchronous Data Communications Equipment*
- EIA Standard RS-366: *Interface Between Data Terminal Equipment and Automatic Calling Equipment for Data Communications*
- EIA Standard RS-404: *Standard for Start-Stop Signal Quality Between Data Terminal Equipment and Non-synchronous Data Communications Equipment*

- **EIA Standard RS-422:** *Electrical Characteristics of Balanced Voltage Digital Interface Circuits*
- **EIA Standard RS-423:** *Electrical Characteristics of Unbalanced Voltage Digital Interface Circuits*
- **EIA Specification RS-449:** *General Purpose 37-Position and 9-Position Interface for Data Terminal Equipment and Data Circuit Terminating Equipment Employing Serial Binary Data Interchange*

3. Another section of the library can be devoted to the "soft side" of the network documentation. Information on architectures, protocols, operation manuals, software manuals, and the like could go here.
4. The last major section in the technical/reference section could be devoted to the "hard side." Technical manuals, schematics, test procedures, etc. would be appropriate here.

When you order equipment, do you specify schematics and maintenance manuals as a line item on the order? This is not a bad habit to get into; you have vendor interest at the time of the order and they are much more willing to part with this material. And don't worry about the confidential schematic ploy; some vendors may make you sign some type of nondisclosure statement. In some cases, it makes you wonder if the purpose is to conceal some of their bad design practices. It's generally worth the hassle, though, to get the technical information which you may need desperately later on.

Data Comm Literature Material

Here is where you stuff all the books and magazines of interest to the network function. I think it's good policy to save back issues of magazines for, say, 3 years. The information in these periodicals is much more up to date than that in almost any other source, and it is often useful to have material available to assist in your studies, proposals, evaluations, etc. Books are useful for more basic and long-term information; you might even stash a copy of this book in the literature section. People in your shop do subscribe to *Data Communications* magazine, don't they? Many people in the data comm business have high regard for this particular periodical.

Archive Library Material

Why save all this stuff in an archive? I'm not suggesting that virtually all your documentation qualifies as archives, but I am suggesting that it's good policy to save some of it. If you have housecleaners in your shop, surely you have run into a situation where you would give your eye teeth for something that was thrown out earlier. You need a compromise. Look over the materials included in your documentation and decide which has a potential of being useful in the future. You might adopt some such guidelines as these for making a decision:

- Does it have basic information which would be useful as general reference material?
- Can it be used as a source of statistics on your network? Many organizations have 5-year plans, for example, which might make good use of some of the network documentation.
- When errors are made in the production documentation and don't surface for weeks or months, the archived material can be a very useful device to track down what happened.

OPERATIONAL DOCUMENTATION

Forms

Everyone seems to have a collection of forms. In some data comm shops, forms are a very necessary item. I am not going to try to make an exhaustive list of possible forms; instead, I'll just concentrate on two basic forms which I think have merit in most networking organizations. One form ties all groups or interested divisions together, and another makes sure that the network staff has a device to help ensure that all tasks necessary to complete a project are done.

How do you ensure that everyone has the same information and that everyone is informed of a new customer coming onto your network, for example? The network staff must worry about the connection of the link; system programmers perhaps need to make sure that the new customer is defined in host software properly; and the finance people need billing information. Most often, not all these people are in the network group, but all must be coordinated. Even if they are in the same group, they need coordination.

A form is the most natural coordinating mechanism. It can be handled either serially or in parallel among the groups concerned. I find that a serial process has the fewest complications, assuming you can get general agreement that it won't sit on someone's desk for too long. In the parallel process not everyone is necessarily looking at the same information. Then there are several forms to control, one for each concerned group. Also, changes in the originally generated form are more difficult to manage when parallel form processing is used. A serially processed form has the drawback that it takes longer for everyone to get it. It is not too difficult to live with this inconvenience, however, in view of the benefits of simplicity and accuracy that go with it. For rush jobs, the form can be hand-carried, if necessary, to help speed up the processing.

Figure 8-1 gives an example of a form intended to be serially processed. It is from a shop having a teleprocessing coordinator who is responsible for control of

```
┌─────────────────────────────────────────────────────────────────────┐
│                    TELEPROCESSING CHANGE REQUEST                     │
│                          (Jan 25, 1982)                              │
│   ACTION:                                                 ID#  ____  │
│    ___ New    ___ Change    ___ Documentation Only   ___ Information │
│                                                                      │
│   USER INFORMATION:              User Target Date: ___/___/__        │
│   User Agency: _____        Contact Person: _____         │
│   Equip. Location: _____  Phone Number: _____         │
│   Proc. Number: ____                                                 │
│                                                                      │
│   PORT AND CIRCUIT DEFINITION:                                       │
│   TCU:   A      B     C     D     E     F     G     H     I    J   K │
│         3272   WSU   OLY   LOC   CNS  BGN1  3274  BGN2 3274          │
│   PORT Addr. ____  Remote Subline ____ Local Subline ____ 470 UCB ___│
│   Host System              CMS Subsc.: Y  N       Line No.           │
│   (CMS/JES2/CICS/TILS/ETC.)  DIAL: MILT  CICS NONE (nn/WLnn/UCSn)    │
│   Circuit No. _____  Terminal Type _____  Data Rate ____   │
│   Near Modem _____     Far Modem _____  Cost Code _____      │
│                                               (Port)                 │
│                                                                      │
│   ROUTING:   (Enter your info on back, sign, & pass on)  Date Passed On: │
│   REQUESTED BY    _____              __/__/__    │
│   TP COORDINATOR  _____              __/__/__    │
│   DATACOMM & ENG. _____              __/__/__    │
│   BRAEGEN         _____              __/__/__    │
│   SYSTEMS         _____              __/__/__    │
│   FINANCE         _____              __/__/__    │
│                                                                      │
│   SUMMARY/COMMENTS:                                                  │
│                                                                      │
└─────────────────────────────────────────────────────────────────────┘
```

(*a*)

Fig. 8-1(*a*) Group coordination form (page 1).

```
┌─────────────────────────────────────────────────────────────────────┐
│                        SCHEDULING & CLOSEOUT                        │
│                                                                     │
│   Datacomm & Eng.:                    Date Scheduled:        By:    │
│                                                                     │
│       Hardware Option = _ _                                         │
│                                                                     │
│       Port                            _ _/_ _/_ _           _____   │
│                                                                     │
│       Circuit                         _ _/_ _/_ _      Engr/Telco/User │
│                                                                     │
│       TCU Modem                       _ _/_ _/_ _      Engr/Telco/User │
│                                                                     │
│       Term. Modem                     _ _/_ _/_ _      Engr/Telco/User │
│                                                                     │
│       Terminal                        _ _/_ _/_ _      Engr/Telco/User │
│                                                                     │
│       _____          _ _/_ _/_ _           _____   │
│                                                                     │
│                                                                     │
│   Systems:                                                          │
│                                                                     │
│       Gen ID: A/B/C/D/E/F/G/H                                       │
│                                                                     │
│       Emulator ID:  _ _ _ _                                         │
│                                                                     │
│       Software Option = _ _                                         │
│                                                                     │
│       Line Set Name                   _ _/_ _/_ _           _____   │
│                                                                     │
│       COMTEN Gens  _ _ _ _ _ _ _      _ _/_ _/_ _           _____   │
│                                                                     │
│       MILTEN/WYLBUR Changes           _ _/_ _/_ _           _____   │
│                                                                     │
│       JES2/HASP Changes               _ _/_ _/_ _           _____   │
│                                                                     │
│       O/S I/O Gen Changes             _ _/_ _/_ _           _____   │
│                                                                     │
│       CICS Changes                    _ _/_ _/_ _           _____   │
│                                                                     │
│       CMS Exec Changes                _ _/_ _/_ _           _____   │
│                                                                     │
│       _____          _ _/_ _/_ _           _____   │
│                                                                     │
│                                                                     │
│   Finance:                                 Equipment/Action being provided │
│       Billing:            Amount:   Proc. No.:  by the user & dates available: │
│                                                                     │
│       Circuit Installation   $ _____  _ _ _ _                     │
│       Port Installation      $ _____  _ _ _ _                     │
│       Terminal Installation  $ _____  _ _ _ _                     │
│       Monthly Circuit Charge $ _____  _ _ _ _                     │
│       Monthly Modem Charge   $ _____  _ _ _ _                     │
│       Monthly Port Charge    $ _____  _ _ _ _                     │
└─────────────────────────────────────────────────────────────────────┘
```

(*b*)

Fig. 8-1(*b*) Group coordination form (page 2).

the form. The first page is filled out principally by the coordinator, and the page serves, in part, as a guide to the coordinator for things to ask of those requesting a change. The second page is for use of the specialists who are responsible for implementing the TP change request. As the form passes through the various groups, information is added to it by those responsible in each group. For that reason the order of flow can be important to help minimize confusion. The people in the finance group, for example, generally do not add information to the form which is needed, say, by the systems people, who precede them. It is impossible to always satisfy everyone on the first pass of the form. This is where the coordinator must take note and clear up missing details if the first pass doesn't get all the information.

If a change in the form is required, a small "change notice" can be attached to the original form and routed back through the various groups to make sure everyone is aware of the change. This ensures that everyone is looking at the information the change refers to. An ID number is assigned to the request by the coordinator. This number is a master reference against which all subsequent documentation can be referenced. The closeout section is useful to help ensure that certain tasks are not overlooked. When complete, the form can be routed through the various groups one last time to let everyone know it is being closed out. This serves as a double check also. As a last comment on this form, the blank spots on the first page which need filling in are keyed to how many characters should be entered for that particular information. This makes it easier for the people who update the documentation from the form.

So much for the form of the first kind. Now, let's look into how the network staff is supposed to keep track of what it is supposed to do when implementing a change to the network. I hope you don't trust the tasks to memory. Some jobs take literally months to complete; you need some mechanism to assist and control this function. Figure 8-2 shows a work-in-progress form taken from a real situation. As shown, the form is intended as a checklist for all the details necessary to implement a circuit. You would have to add sections to or subtract them from the form to suit your own situation. It wouldn't be a bad idea for other groups in the organization to have a similar control to help ensure that their respective responsibilities are not overlooked. The idea here is to ensure that when a request comes through your shop (via the TP change request form or its equivalent), it gets task assignments to implement it (via the work-in-progress form). In practice, you might make a copy of the TP change request form as it comes by, attach it to a work-in-progress form, and file it in the data comm records. That way you will have a full record of that particular job. The form is fairly self-explanatory; in essence, it is simply a checklist of things needing attention to make a change to the network. Some other points to note are:

- Use this form in conjunction with the TP change request form; the ID number should be the number assigned to the TP change request form to link the two forms together.

DOCUMENTATION

```
┌─────────────────────────────────────────────────────────────────────┐
│              DATA COMM WORK-IN-PROGRESS            ID#_____        │
│                                                                     │
│  Computer End Work:                                                 │
│   ┌───────────────────────────────────────────────────────────────┐ │
│   │ EQUIP INSTALL:    CABLING:          LABELING:                 │ │
│   │ ___FEP Port Hdwr  ___FEP Port       ___CDC                    │ │
│   │ ___327X Port Hdwr ___327X Port      ___LDC                    │ │
│   │ ___Diag. Port Hdw ___Diag. Port     ___Modem Shelf/Rack       │ │
│   │ ___Modem:         ___LDC Line       ___Loopback unit          │ │
│   │ ___Other:         ___Modem Line     ___Diag. Port             │ │
│   │                   ___Other:         ___Ownership Tags         │ │
│   │                                     ___Other:                 │ │
│   └───────────────────────────────────────────────────────────────┘ │
│                                                                     │
│  Terminal End Work:                                                 │
│   ┌───────────────────────────────────────────────────────────────┐ │
│   │ EQUIP INSTALL:    CABLING:          LABELING:                 │ │
│   │ ___Modem          ___Terminal Drop  ___Circuit Number         │ │
│   │ ___Terminal       ___Modem Cabling  ___Ownership Tags         │ │
│   │ ___Other:         ___Other:         ___Other:                 │ │
│   └───────────────────────────────────────────────────────────────┘ │
│                                                                     │
│  Documentation Updates:                                             │
│   ┌───────────────────────────────────────────────────────────────┐ │
│   │ GENERAL:          WSU FILES:   OLY FILES:     MISC. FILES:    │ │
│   │ ___Sched. Calendar ___WSUCDC   ___OLYCRKT     ___FEP Port List│ │
│   │ ___Cab & Adap Book ___WSUDIAL  ___OLYFDMS     ___327X Port List│
│   │ ___Modem Ref Book  ___WSUEQ    ___OLYLABEL    ___CUSTEQ       │ │
│   │ ___Other:          ___WSUFDMS  ___OLYTEL#     ___Terminal File│ │
│   │                    ___WSULABEL ___OLY Bluelines ___Coupler File│
│   │                    ___WSULDC   ___Other:      ___Other:       │ │
│   │                    ___WSULONG                                 │ │
│   │                    ___WSUSHRT                                 │ │
│   │                    ___WSUTEL#                                 │ │
│   │                    ___Other:                                  │ │
│   └───────────────────────────────────────────────────────────────┘ │
│                                                                     │
│  Work Description/Comments:                                         │
│   ┌───────────────────────────────────────────────────────────────┐ │
│   │                                                               │ │
│   │                                                               │ │
│   │                                                               │ │
│   │                                                               │ │
│   └───────────────────────────────────────────────────────────────┘ │
│                                                                     │
│  Dates:                          Closeout:                          │
│                                                                     │
│  Telco Release    : _____    ___Circuit Number Documented?     │
│  _ _ _ _ _ _ _ _  : _____    ___Port Listing Updated?          │
│                   : _____    ___Replacement Parts Ordered?     │
│  Date Completed   : _____    _ _ _ _ _ _ _ _ _ _ _ _ _ _       │
└─────────────────────────────────────────────────────────────────────┘
```

Fig. 8-2 Work-in-progress form.

- Another device useful with this form is a calendar showing target dates of projects at a glance. The ID number on the calendar refers one back to the work-in-progress form for more detailed information or to the TP change request form for other general information.

- Use the closeout section as a double check on certain items.
- One frequently overlooked item is to reorder stock used to implement a project.

Records

Quite a variety of records can be necessary to accomplish the data comm function. Trouble logs, maintenance logs, performance stats, equipment inventories, and preventive maintenance logs are examples. The following are a few points to consider in regard to these records:

- If your system is shut down over the weekends, do you keep a log of the maintenance and other activity that transpires during the time you are not in production? Come Monday morning and you can't get the system running, it may be because of some work done during nonproduction hours. A log will help pinpoint activities which may have a bearing upon the problem.
- Do you include documentation in your PM program?
- Do you use the records you generate? Sometimes record collecting becomes such a habit that people don't realize they never really use what they have collected.

Miscellaneous Operational Documentation

Included in miscellaneous documentation are things like operator's reference manuals and telephone contact lists. There is not much to say about these items except that they are an important part of the overall operational documentation. Does your trouble call-out list show each piece of equipment or function and who is responsible for its maintenance? Is a backup listed also?

NETWORK DOCUMENTATION*

On larger computer systems, some form of network management system can take care of a lot of the network documentation. In the smaller shops, however, there is still need to worry about this documentation and implement it efficiently. Here we are concerned with documenting the network from the port out to the end user of your system. In this section, I will stay at a basic level and primarily address the newcomers and the smaller network documentation needs. No doubt the larger network shops have more elaborate documentation efforts tailored to their particular situations.

* This section, up to and including "Coaxial Circuit Documentation," is excerpted from *Application Notes 6, 7,* and *8,* © copyright 1981 by Computer Energy, Inc. (P.O. Box 2096 CS, Pullman, WA 99163).

The first requirement of any computer network documentation function should be to perform the documentation via a computer file. Any system large enough to have an associated network surely also has text-editing facilities to allow efficient documentation procedures. This section assumes that this requirement has been met as a basis for developing a set of procedures and conventions useful in network documentation. It should be pointed out that even systems having some type of network management/diagnostic system software should still use some form of hard copy for documentation of circuit information. It must also be stressed that we are considering only the documentation necessary to trace the circuit path, not all the additional information about the system port to which it is connected.

The circuit documentation file is necessarily accessed and updated by several people within an organization. Because of this, it is important that procedures be established to control how updates are performed so that consistency and understandability are maintained. The following procedures are advisable:

1. No autonomous decisions should be allowed. Changes made arbitrarily by one person lead to confusion among all those who are responsible for the circuit documentation and its use. Documentation changes should involve and be approved by everyone concerned (or perhaps only the manager for networking) before they are allowed to be implemented.

2. Two files are useful: the "just updated" file and the file as it existed just before the update. Call these NEW and OLD, respectively. They should be used as follows to perform an update to the circuit documentation:

    Step 1: Fetch a copy of the OLD file and perform the necessary updates.

    Step 2: Save the updated file under NEW.

    Step 3: Only when sure that NEW is accurate, replace OLD with NEW. This generally happens when the hard copy of NEW has been examined.

    Following the above procedure will save much time and frustration when a file is inadvertently destroyed. It also permits easy tracking and correction of simple documentation errors, which always seem to occur.

3. It is useful to insist that only one "working" hard copy of the circuit documentation exist and that it be located within the network area, where it is accessible to everyone concerned. Further, insist that changes written on the working hard copy be made in high-contrast ink, such as red ink, for immediate identification of circuit changes. Following this procedure will prevent the confusion that would arise from multiple copies (and revisions). It is permissible for people to have their own copies, but insist that decisions regarding or reference to network information be done via the working hard copy. Updates to the circuit documentation file on the computer should be done only by referencing the working hard copy with the update changes flagged in red ink.

138 DOCUMENTATION

4. An archive should be established and a decision made on how long to archive the outdated circuit documentation. Three to five years is generally a reasonable time period. Archiving network documentation allows for later analysis such as network projections and utilization trends.

Certain information should be included with the specific circuit parameters to assist in the maintenance of the file. This control information varies among networks, but consideration should be given to at least the following:

1. The file name should be included so that it shows on the hard copy for reference.
2. The date should be included. Dating done 2 JAN 83, for instance, is less confusing and less prone to error than 2/1/83. Or is it 1/2/83?
3. Include routing information as to who outside the network staff should receive copies.
4. A title block or banner which readily identifies the document should be included.
5. Include any special points which apply to the document as a whole. For instance, in-house circuits *not* using modems require a cross of the SEND and RECV pairs. You may want to make this clear in the circuit documentation.
6. Column numbering is useful for easing the task of file updating with line editors.

It is important not to overstuff the circuit file with overhead information. There is generally too much information regarding each network line to put in one document effectively. In practice, several documents will exist; the circuit document will be just one of them. It is important to decide on points which link the various documents together and then include only relevant information in each document. In the case of the circuit document, its purpose is to document the circuit path within the responsibility of the network staff. This may overlap somewhat with telephone company circuit responsibilities. The circuit documentation should cover the path up to but not including the port. The port documentation is generally on a separate document, and its link to the circuit documentation should be via the circuit number (or telephone number) listed on the port document itself.

Twisted Pair Circuit Documentation

Of concern here is how to document the circuit itself—how it routes from one end to the other and the linkage points in between. Twisted pair and coaxial circuits are discussed separately because of the differences in their implementation. It is more practical to have two individual documentation efforts for these media rather than devise a scheme to document both together.

Information regarding the network varies among installations depending upon services offered and types of equipment in use. For twisted pair circuits, the following categories generally exist:

1. Telco dial circuits
2. Telco short-haul circuits
3. Telco long-haul circuits
4. In-house local active circuits
5. In-house local inactive circuits

Depending upon the size of the network, these five categories may exist in one to five separate documentation files. It is generally preferable to have them under individual files within the data base. Common to the five categories are two minimum documentation requirements:

1. Circuit number or telephone number
2. Wiring path documentation

Additional supplementary information can be added to the above basic information:

1. Terminal end user information such as location and name
2. Location information regarding circuit components such as the shelf or rack location of circuit modems
3. Circuit component information such as modem type
4. Comment field for special circuit notes
5. Blank lines to write in new circuit information

The suggestions made above are summarized in Figures 8-3(a) through 8-3(e), which show examples of what the hard copy might look like for the five general circuit categories mentioned.

Coaxial Circuit Documentation

The scope of this section is limited to 327X coaxial networks, although some of the ideas may apply to other coaxial circuits. Of interest here is documenting the coaxial circuit from the controller out to the attached 327X device. For remote controllers, the documentation is identical. A note on the remote documentation can show the Telco circuit number, which is the only information needed to reference someone to the twisted pair documentation covering the Telco portion of the circuit.

Documentation of 327X circuits needs to be concerned with the following categories of circuits:

```
FILENAME:   DIALS
DATE:       1 MAY 81
COPIES:     FINANCE

                    ****************************************
                    *                                      *
                    *                                      *
                    *         D I A L   D A T A   C I R C U I T S         *
                    *                                      *
                    *                                      *
                    ****************************************

        1       2       3       4       5
        1       0       2       4       8
TEL.    CABLE   NEXT    LAST
NUMBER  PAIR BL BLOCK   BLOCK   MODEM   COMMENT
------- ------- ------- ------- ------- -------
123-1000 9-8765                 113-A-01  113D    300 BAUD ASYNC, TEST PORT
123-1100 9-8764 MDM1-A-10 MDM1-A1-10 202S 1200 BAUD ASYNC
123-2000 9-8763 MDM2-A-20 MDM2-A1-20 201C 2400 BAUD SYNC
123-2100 9-8762 MDM2-A-30 MDM2-A1-30 208B 4800 BAUD SYNC
123-3001 9-8700                 113-A-10  113B    123-3001 ROTARY
123-3002 9-8701                 113-A-11  113B    123-3001 ROTARY
123-3003 9-8702                 113-A-12  113B    123-3001 ROTARY
123-3004 9-8703                 113-A-13  113B    123-3001 ROTARY
```

Fig. 8-3(*a*) Dial circuit documentation example.

FILENAME: TELSHRT
DATE: 1 MAY 81
COPIES: FINANCE

```
***********************************************************
*                                                         *
*                                                         *
*    T E L C O   S H O R T   H A U L   C I R C U I T S    *
*                                                         *
*                                                         *
***********************************************************
```

1	11	17	29	38	50	62	73	87	105
CIRCUIT NUMBER	PAIR	REMOTE BLOCK	CABLE PAIR	NEXT BLOCK	LAST BLOCK	MODEM LOCATION	MODEM	CUSTOMER LOCATION	COMMENTS
999-0001	RECV SEND		6-2375 6-2376	PRN-G1-01 PRN-G1-02	PRN-G-01 PRN-G-02	PRN-G1	ASYNC LDM	BLDG A, ROOM 100	
999-0025	RECV SEND		6-2239 6-2240	MDM3-38 MDM3-39	MDM3-B-38 MDM3-B-39		SYNC LDM	BLDG B, ROOM 200	RJE STATION
999-0079	RECV SEND	TLAB-A-32 TLAB-A-33	7-0001 7-0002	PRN-E1-01 PRN-E1-02	PRN-E-01 PRN-E-02	PRN-E1	ASYNC LDM	BLDG C LAB	
999-0080	RECV SEND	TLAB-A-09 TLAB-A-10	7-0003 7-0004	PRN-E1-03 PRN-E1-04	PRN-E-03 PRN-E-04	PRN-E2	ASYNC LDM	BLDG C LAB	
999-0327	RECV SEND		6-2307 6-2308	PRN-A1-03 PRN-A1-04	PRN-A-03 PRN-A-04	PRN-A2	ASYNC LDM	BLDG D, ROOM 400	
	RECV SEND								NEW CIRCUIT
	RECV SEND								NEW CIRCUIT

Fig. 8-3(*b*) Short-haul circuit documentation example.

```
FILENAME:  TELLONG
DATE:      1 MAY 81
COPIES:    FINANCE

            ********************************************
            **                                        **
            **        T E L C O   L O N G   H A U L   C I R C U I T S        **
            **                                        **
            **                                        **
            ********************************************
```

	1 7	2 2	3 0	3 6	4 2	4 8	5 9	7 0	8 0	9 3	1 0 7
CIRCUIT NUMBER	FAIR	CABLE PAIR	: 412 EQUIPMENT :BLOCK CABLE MODEM : SIDE SIDE		:NEXT :BLOCK :		LAST BLOCK	MODEM LOCATION	MODEM	CUSTOMER LOCATION	COMMENT
FDEC-999031	RECV SEND	5-2448 5-2450	:LB2 :LB2	-07 -08	:-32 :-33	:MDM4-26 :MDM4-27	MDM4-B-26 MDM4-B-27		2024	NEBRASKA	MULTI-DROP
FDDC-99967	RECV SEND	5-2317 5-2318	:LB4 :LB4	-07 -08	:-32 :-33	:MDM4-48 :MDM4-49	MDM4-B-48 MDM4-B-49		48B1	BOISE	
FDLC-99967-003	RECV SEND	5-2311 5-2312	:LB4 :LB4	-21 -22	:-46 :-47	:MDM2-32 :MDM2-33	MDM2-B-32 MDM2-B-33		208A	ALASKA	
FDLC-99967-004	RECV SEND	5-2371 5-2230	:LB3 :LB2	-03 -24	:-28 :-49	:MDM1-30 :MDM4-06	MDM1-B-30 MDM4-A-06		208A	PHOENIX	
	RECV SEND										NEW CIRCUIT
	RECV SEND										NEW CIRCUIT

Fig. 8-3(c) Long-haul circuit documentation example.

```
 1.
 2.  FILENAME:    LDC
 3.  DATE:        1 MAY 81
 4.  COPIES:      NONE
 5.
 6.
 7.       ************************************************
 8.       *                                              *
 9.       *                                              *
10.       *            L O C A L   D A T A   C I R C U I T S   *
11.       *                                              *
12.       *                                              *
13.       ************************************************
14.
15.
16.
17.       ****************************************
18.       *                                      *
19.       * IMPORTANT:  THE CROSS BETWEEN SEND AND RECV OCCURS *
20.       *             ON THE LAST BLOCK AT THE TERMINAL END  *
21.       *             WHERE THE TERMINAL WIRES ARE CONNECTED *
22.       *             TO THE BLOCK.                          *
23.       *                                      *
24.       ****************************************
25.
26.
27.
28.                                                                    1
29.           1       1                 3         4         6    7    8    9 0 3
30.  1       0       6                 1         6         0    4    6    9
31.
32.  CIRCUIT         PAIR    FIRST     NEXT      NEXT      NEXT NEXT LAST TERMINAL LOCATION
33.  NUMBER                  BLOCK(S)  BLOCK(S)  BLOCK(S)  BLOCK(S) BLOCK BLOCK    COMMENT
34.  -------         ----    --------  --------  --------  -------- ----- -----    ----------------
35.
36.  LDC-A01         RECV    LDC-A/A1-01    CPT2-D-01                         CPT2-D1-01  R:S  CPT 2147B
37.                  SEND         -02           -02                              -02      S:R  OPERATIONS
38.
39.  LDC-A02         RECV    LDC-A/A1-03    CPT2-D-47                         CPT2-D1-47  R:S  CPT 2147A
40.                  SEND         -04           -48                              -48      S:R  SYSTEMS
```

Fig. 8-3(d) In-house active circuit documentation example.

143

```
FILENAME:      NOPORT
DATE:          1 MAY 81
COPIES:        NONE

****************************************************
*                                                  *
*                                                  *
*          I N A C T I V E   D A T A   C I R C U I T S
*                                                  *
*                                                  *
****************************************************

       ****************************************
       *                                      *
       *   IMPORTANT:  THE CROSS BETWEEN SEND AND RECV OCCURS   *
       *              ON THE LAST BLOCK AT THE TERMINAL END     *
       *              WHERE THE TERMINAL WIRES ARE CONNECTED    *
       *              TO THE LAST BLOCK.                        *
       *                                      *
       ****************************************


                                                                              1  1
                          1            2       4         5        6        8  0  9
         1    1           8            8       0         4        8        0  5

CIRCUIT       PAIR  FIRST        NEXT         NEXT     NEXT     NEXT     NEXT     LAST       DROP
NUMBER              BLOCK        BLOCK        BLOCK    BLOCK    BLOCK    BLOCK    BLOCK      LOCATION
------  ----  -----------  -----------  -------  -------  -------  -------  -----------   -----------
  1

LDC-001       RECV  MDM4-A-07    MDM4-07      CPT2-A-45                            CPT2-A1-45  R:S    CPT 2176
              SEND       -08         -08           -46                                   -46  S:R

LDC-002       RECV  CPT2-L-05                                                      CPT2-L1-05  R:S    CPT 2001
              SEND       -06                                                             -06  S:R

LDC-003       RECV  CPT2-D-26                                                      CPT2-D1-26  R:S    CPT 2138
              SEND       -27                                                             -27  S:R
```

Fig. 8-3(e) In-house inactive circuit documentation example.

1. Point-to-point active circuits
2. Daisy-chained active circuits
3. Inactive circuits

One document can be used to list active coaxial circuits of both types. Inactive circuits should be documented to keep track of the unused coaxial cable and should be documented separately from the active circuits. The only real difference between the active and inactive types of documentation is the lack of a cable assignment to the controller for the inactive circuit.

Coaxial circuit documentation must show how the circuit actually exists from the port to the device. Where large numbers of coaxial cables are in use, they may be installed in bundles or feeders, and the documentation must allow for this. A last point to consider is the definition of a coaxial data circuit (CDC). When you assign an arbitrary circuit number to lengths of coax between a terminal and a controller, what exactly does the number apply to as viewed from the port end? When the CDC is compared with the local data circuit (LDC) described in an earlier chapter, it is noted that there is generally a single cable connecting a port to an LDC, whereas there may be a multiple-cable feeder connecting a 327X controller to a coax patch panel. This difference may influence how the CDC documentation is formatted.

The preceding discussion is illustrated in Figure 8-4, which shows how CDC documentation can be implemented.

Port Documentation

The network staff needs access to certain port-related information to perform the data comm function efficiently in DP service situations. The port into a computer system is a reasonable central reference point for linking a variety of documen-documentation. The port links a data circuit to a service and, in turn, a particular user to the system in many cases. The port may also represent a particular class of service such as asynchronous dial access versus synchronous dial access.

The port document can have a large variety of information relevant to several different functions and people. Getting all the information on one document may be an impossible task. Figures 8-5 and 8-6 give examples of port documents with information mainly of concern to the data comm staff.

The port documentation can supply most of the information regarding who is connected and where, what systems are being used, what UCBs are allocated, etc. A point I would like to bring out here is that very likely more than one organization or group uses (or could use) this document, but each user group of the document wants a little something different from the other group and would prefer not to have its document cluttered with all that extraneous information used by other people. If you have a program similar to "Spires," why don't you consider using it to solve this problem. Create one large file with categories of

```
1.  FILENAME:    CDC
2.  DATE:        1 MAY 81
3.  COPIES:      NONE
4.
5.
6.
7.          ****************************************************
8.          **                                                **
9.          **              COAXIAL DATA CIRCUITS             **
10.         **                                                **
11.         ****************************************************
12.
13.
14.
15.
16.
17.          1        2          4       5         7         8
18.          7        9          4       7         0         6
19.     CONTROLLER  CIRCUIT   NEXT    NEXT    LAST      TERMINAL  TERM. DROP
20.     FEEDER     NUMBER    FEEDER  FEEDER  FEEDER    DROP      LOCATION   COMMENTS
21.     -------    -------   ------  ------  -------   --------  ---------- --------
22.
23.     FDR1-012-07  CDC-A01  FDRB-001-01            BB-004-01              BLDG 1, RM A
24.
25.     FDR1-012-16  CDC-A02  FDRB-001-02            BB-004-02              BLDG 1, RM B
26.
27.     FDR1-012-08  CDC-A03  FDRB-001-03            BB-001-01              BLDG 2, RM A
28.
29.
30.
31.     FDR1-013-03  CDC-B01  FDRB-001-07            BB-002-01   CDC-B01-A  BLDG 3, RM A  START OF B01 CHAIN
32.
33.                                                              CDC-B01-B  BLDG 3, RM B
34.
35.                                                              CDC-B01-C  BLDG 3, RM C
36.
37.                                                                         BLDG 3, RM D
38.
39.     FDR1-013-10  CDC-B02  FDRB-001-05            BB-003-01   CDC-B02-A  BLDG 4, RM A  START OF B02 CHAIN
40.
41.
42.
43.     -----------
44.
45.     FDR1-011-04  CDC-C01                                                I/O RM        DIRECT - TAPE CONSOLE
46.
47.     FDR1-011-03  CDC-C02                                                I/O RM        DIRECT - MAIN CONSOLE
48.
49.     FDR1-011-11  CDC-C03                                                I/O RM        DIRECT - 3286 HRDCPY
50.
51.     FDR1-011-07  CDC-C04                                                DATA COMM     DIRECT - DATA COMM
52.
```

Fig. 8-4 Coaxial circuit documentation example.

```
 1. FILENAME: COMPORT
 2. DATE:     21 AUG 81
 3. COPIES:   FINANCE, SYSTEMS
 4.
 5. ****************************************
 6. *                                      *
 7. *        'E' FRONT-END PORTS           *
 8. *                                      *
 9. ****************************************
10.
11.
12. GEN DATE: 15 AUG 81                                              PAGE 1
13.
14.          1    1                    2   2   3   3   4   4
15.    1  6  0    5                    3   7   3   8   3   8        6   7                   8   7             1
16.                                                                 2   1                   7   7             0   7
17. ADDRESS TP  SUBLINE 470 OPTS HOST LINE DATA CIRCUIT   TERMINAL                          NEAR     FAR
18. HEX DEC RGST LDC RMT UCB HW SW SYS ID  RATE NUMBER    TYPE      LOCATION        USER    MODEM    MODEM
19. ------------------------------------------------------------------------------------------------------
20.
21. E000  0                        30 00 CNS   T1  9600  4KD-0525-01 TRUNK1       PULLMAN         WSUCSC   SFD-LSI96 SFD-LSI96
22. E001  1                        30 00 CNS       9600                                           WSUCSC
23. E002  2                        30 00 CNS   T2  9600  4KD-0525-02 TRUNK2       PULLMAN         WSUCSC   SFD-LSI96 SFD-LSI96
24. E003  3                        30 00 CNS       9600                                           WSUCSC
25. E004  4    064                 30 00 HASP  8   4800  950-0025   REM-4780     WSU CARPENTER    UCS      SGE-330   SGE-330
26. E005  5    0B3                 00 00 HASP 25   2400  TEST PORT  REM-2780     TEST PORT        TEST PORT
27. E006  6    062                 00 01 HASP  6   2400  335-2561   HASP-RJE     WSU (DIAL)       VARIES   STEL-201C STEL-201C
28. E007  7    0CB                 00 00 CICS      4800                                           SPARE
29. E008  8    07F                 00 00 CICS      4800                                           WSUCSC
30. E009  9    0C9                 00 00 CICS  31  4800  LDC-G04                 IBM 6670  WSU CSB 2176    SPARE
31. E00A 10    06A 80              00 00 HASP                                                     WSUCSC
32. E00B 11    06C                 30 00 HASP  4   4800  208-0002   IBM 360/     WSU CSB 2176    OPERATOR SGE-330   SGE-330
33. E00C 12    070                 00 00 HASP 17   2400  208-0002   PRINTRNX    USFS-MOSCOW      USFS     STEL-201C STEL-201C
34. E00D 13    06E                 00 00 HASP 13   2400  4KD-0376   REM-2780    MOSES LAKE       BRCC     SULT-2400 SULT-2400
35. E00E 14    0B4                 00 00 HASP 26   2400  950-0111   PDP 15       MCCOY S206      VET MED  SPRN-SLD  SPRN-SLD
36. E00F 15    0CA                 00 00 CICS      4800                                           SPARE
37.
38.
39. E040 64                         00 00 TILS      1200  LDC-006    TVI-912C    WSU CSB 2047    CPT SCI
40. E041 65                         00 00 TILS      1200  LDC-N03    HAZ-MOD1    WSU CSB 3089    UCS
41. E042 66 815C                    00 00 TILS      1200             FOX-1100    DANA 136        WSU LAB
42. E043 67                         00 00 TILS      1200  LDC-N04    TVI-912C    WSU CSB 3043    UCS/ALMNI  AFRN-ALD AFRN-ALD
43. E044 68                         00 00 TILS      1200  LDC-N05    TVI-912C    WSU CSB 3025    UCS/ALMNI
44. E045 69                         00 00 TILS      1200  LDC-N06    TVI-912C    WSU CSB 3033    UCS/ALMNI
45. E046 70                         00 00 TILS      1200  LDC-N07    TVI-912C    WSU CSB 3047    UCS
46. E047 71                         00 00 TILS      1200  950-0343   TVI-912C    JTOWER 520      BUS ADM    AFRN-ALD AFRN-ALD
47. E048 72                         00 00 TILS      1200  LDC-N08    TVI-912C    WSU CSB 3037    UCS
48. E049 73                         00 00 TILS      1200  LDC-N09    TVI-912C    WSU CSB 3041    AMDAHL
49. E04A 74                         00 00 TILS      1200  LDC-H08    TVI-912C    WSU CSB 1021    UCS
50. E04B 75                         00 00 TILS      1200  LDC-N10    TVI-912C    WSU CSB 3049    UCS
51. E04C 76                         00 00 TILS      1200  LDC-N11    TVI-912C    WSU CSB 3083    USER SVC
52. E04D 77 887                     00 00 TILS      1200             FOX-1100    MOSCOW          USFS       ATMX-MUX ATMX-MUX
53. E04E 78 895                     00 00 TILS      1200             TEK-4010    MOSCOW          USFS
54. E04F 79 896                     00 00 TILS      1200             LA-120      MOSCOW                     ATMX-MUX ATMX-MUX
```

Fig. 8-5 Twisted pair port documentation.

```
 1.   FILENAME:  PORT327X
 2.   DATE:      21 AUG 81
 3.   COPIES:    FINANCE, SYSTEMS
 4.
 5.
 6.   ****************************************
 7.   *                                      *
 8.   *          BRAEGEN 327X PORTS          *
 9.   *                                      *
10.   ****************************************
11.
12.             1       2        3       4        5
13.             8       1        2       5        8
14.
15.   PORT  UCB   TP     CIRCUIT    TERMINAL   LOCATION   USER
16.              RQST    NUMBER     TYPE
17.
18.   FOA1  F22   572    CDC-H02    L3277      CSB 3038   UCS
19.         F23                     L3277      CSB 3038   UCS
20.         F26                     L3277      CSB 3064   UCS
21.
22.
23.   F1A1  F28          CDC-A04    L3277      FRN 442    FIN. AID
24.         F29                     L3277      FRN 444    FIN. AID
25.         F14                     L3277      FRN 442B   FIN. AID
26.         F10                     L1403      FRN 442B   FIN. AID
27.
28.
29.   ****************************************
30.   *                                      *
31.   *           IBM 327X PORTS             *
32.   *                                      *
33.   ****************************************
34.
35.
36.   PORT  UCB   TP     CIRCUIT    TERMINAL   LOCATION    USER
37.              RQST    NUMBER     TYPE
38.
39.   A00   E20          CDC-D10    3277       CSB 2176    MAIN CONSOLE
40.   A01   E21   410    CDC-D11    3277       CSB 2176    I/O CONSOLE
41.   A02   E22          CDC-D07    3277       CSB 2176    TAPE CONSOLE
42.   A03   E23          CDC-D16    3277       CSB 1010    DATA COMM
43.   A04   E24          SPARE                 SPARE       SPARE
44.   A05   E25          SPARE                 SPARE       SPARE
45.   A06   E26          SPARE                 SPARE       SPARE
46.   A07   E27          SPARE                 SPARE       SPARE
47.
48.
49.   G01   E40   862    CDC-G09    3279       AVERY 324   HRC
50.   G02   E41          CDC-B01    3279       CSB 2005    SR
51.   G03   E42          SPARE                 SPARE       SPARE
52.   G04   E43          SPARE                 SPARE       SPARE
53.   G05   E44   848    CDC-G11    3278       CSB 2084    CONSULTANT
54.   G06   E45   813A   CDC-D02    3278       CUB 77      ALUMNI
55.   G07   E46          CDC-G01    3278       CSB 2163    SYSTEMS
56.   G08   E47          CDC-G12    3279       CSB 2004    USER SERVICES
```

Fig. 8-6 327X port documentation.

interest to all users, and then print off only that information useful to each of the users. A useful column item to include on the port document is the TP change request ID number which was responsible for the port change. This can be quite useful in tracking down documentation problems, misunderstandings, etc.

Network Feeder Documentation

What should your book documenting all that feeder cable you installed look like? You do have a book, don't you?! Don't feel bad, many others trust their memories too. The book describing the bulk wiring of the facility would, first of all, hopefully include a few procedures and information for new people who are not acquainted with the network. Some subjects to include might be these:

- General wiring procedures for connecting feeders to blocks
- Color-coding procedures
- Sample cable labels
- Philosophy of twisted pair circuits and coaxial data circuits
- A list of mnemonics in use within the book

A book like this will solve some of the personnel turnover problems that arise when information leaves with the departing staff. Past the boilerplate, you will need to include a section where the individual feeder cables can be inventoried by cable number. Figure 8-7 shows an example of such an inventory sheet, which is self-explanatory. One last useful feature of a book like this is to include building diagrams with the cable runs inked in. The diagrams can sometimes be obtained from the architect or others in charge of the facility. Small drawings that often are made by space-planning analysts are ideal for this purpose. This book is also a good place to document conduit lengths for future reference.

MISCELLANEOUS DOCUMENTATION

Labeling

Here are a few comments in regard to labeling which you might find of interest and maybe even of use:

- *Cable Labels*

    Let's start with all those labels you stick on cables. Do you just type on the circuit number and then stick the label on the cable? Here's an improvement: type the number on at least twice so that it is more readable without twisting the cable out of shape and just in case one typing is not too legible. Another thing you can do is underline the last number typed. This will help avoid

BULK FEEDER INVENTORY

Series: <u>CSB1</u>

Cable Number	From	To	Cable Type and Length	Comments
CSB1-001	Data Comm Area	Computer Room	50 pair 100 feet	
CSB1-002	Data Comm Area	Hallway-1080S	100 pair 150 feet	Main feeder to first floor offices
CSB1-003	Data Comm Area	Data Comm Area	50 pair 50 feet	Terminal Block for Data Comm area
⋮				
CSB1-008	Hallway-1080S	Hallway-1061	50 pair 150 feet	First floor office sub-feeder
⋮				
CSB1-020	327X Contr. #1	CDC Racks	(16) RG 62A/U 100 feet	
CSB1-021	327X Contr. #2	CDC Racks	(32) RG 62A/U 100 feet	
CSB1-022	CDC Racks	First floor offices	(8) RG 62A/U 200 feet	Main feeder to 1st floor offices

Fig. 8-7 Network feeder documentation.

confusion in situations where you cannot readily identify which way the label is supposed to be read, e.g., is it 66 or 99?

- *Shelf and Rack Labels*

Size is a problem when you try to put a label in a label holder which is the same height as the thickness of the shelf. If you have many labels at all, they should be on a computer file. The problem is to find a way to do the hard copy correctly. You can get three lines of information with perhaps two columns, which gives at least six pieces of information you can include on the shelf label. This results in a label quite acceptable for most shelf labeling needs but unworkable for tightly packed rack labels. A length conflict also occurs when you try to put several labels below something like a stat mux, which has several port connections to identify. In this case and other cases where you need more data on the label, you might have to resort to copying your hard copy on a machine which can reduce the size. Because of the varying size requirements, you might find it useful to partition your label file into two sections: one for original-size labels and the other for reduced-size labels. Each section of the label file can then be adjusted so the hard-copy output is conveniently sized for rapid shearing of the labels.

- *Equipment Labels*

  It has been my experience that it is best to avoid any labels on equipment if at all possible. When equipment is changed or swapped out during maintenance, you add to your work if labels also have to be transferred. In a rush to solve a problem, you might forget to put a label on the new equipment when you do a swap. This will cause problems later on when you need the information again. Some exceptions to this rule would be labels to identify ownership and those on multiple larger machines, like FEPs, where you need to make it clear which machine is which.

Blueprinting

For documenting a wide range of things which change periodically, you should consider getting into blueprints. They can save you a lot of time and increase the accuracy of this type of documentation. Your volume will be too low to justify equipment for this purpose, but in most any locality there are blueprint machines which you can no doubt arrange to use. Once you have that problem solved, there aren't any other out-of-the-ordinary expenses; you need a few drawing implements even if you don't use blueprints. Some of the things you can document via blueprints are:

- Electrical and/or mechanical things you construct
- Equipment layout and planning sheets
- Facility layout information
- Planning sheets for your subfloor power grid
- Planning sheets for your punch block wiring panel system
- Planning sheets for your local data circuit equipment
- Equipment rack layout for planning high-density equipment
- Shelving layouts and planning of their usage
- Layout of your cable hanger grid system and its utilization

With blueprints it is easy to generate new copies; all you do is correct only that part of the master which needs correction and then run off some new copies. This type of documentation also works better when large sheets of paper are more preferable. Large paper is not easy to use with common office copy equipment, but it is a natural application for blueprints.

Before leaving this subject, I would like to concentrate on an area most likely to be out of most readers' experience: documenting electrical/mechanical things via blueprints. There are things you should consider in this type documentation to increase its utility. The adapter assemblies described in an earlier chapter are an ideal example. Figure 8-8 shows a reduced copy of an adapter assembly blueprint. You should have at least the following information on such blueprints:

1. A mechanical drawing of the device showing how it is constructed
2. An electrical schematic showing how the device functions electrically
3. A short description of the thing's purpose in life, if it is not obvious
4. Any other information which assists in the construction or understanding of the device
5. Documentation aids such as what to label the device and where to label it for consistency

Note in Figure 8-8 that you don't need to have everything ruled or professionally drawn and lettered; that would take away from the utility of blueprints. Hand printing is quite acceptable as long as it is legible. It would be a mistake to go overboard and rule and letter everything unless you need an excuse to create another bureaucracy. Just insist that the blueprints follow some of the above standards for content and that they are reasonably presentable. If you want to dress up your blueprints a little, you can have preprinted labels made like the one shown in Figure 8-8. A reference is given in the Appendix for Bishop Graphics, Inc., who can supply them.

Power Documentation

Putting labels on breakers in the power panels would be the minimum acceptable documentation effort for the data comm facility power. However, additional documentation can enhance your operation:

- It's useful to know how power flows from panel to panel. Having this visible on documentation helps you plan power utilization and changes. In an emergency, would you know where to go to disable a whole power panel?
- The breaker labels are necessarily small. Supplemental documentation allows you to add to the amount of breaker documentation, thereby reducing possible errors and confusion.
- Documenting the type of receptacle connected to the breaker avoids having to lift floor tiles and wade through the wiring to find out.

Figure 8-9 shows how you can go about documenting your power. The example shown has a banner indicating documentation for protected power, i.e., from an uninterruptible power system (UPS). If you run equipment off standard utility power, you might want a separate document for that also. If you put a header before each panel being documented, you can put a note in it to indicate where power to that panel originates, which can be a useful feature. Note also that the

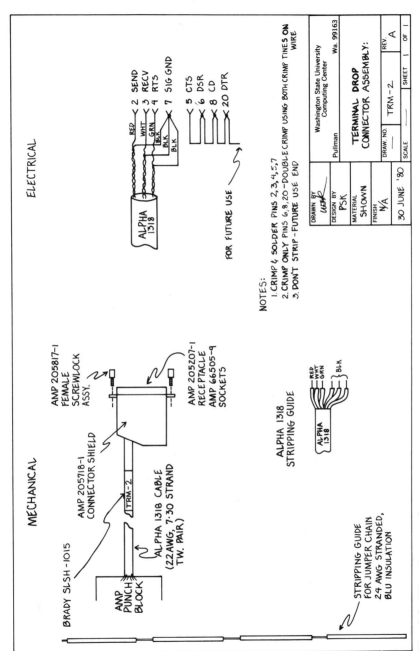

Fig. 8-8 Blueprint example.

154 DOCUMENTATION

```
1.    FILENAME:      #BREAKER
2.    DATE:          26 MAR 81
3.    COPIES:        NONE
4.
5.
6.
7.         ***********************************************************
8.         ***********************************************************
9.         **                                                       **
10.        **                                                       **
11.        **       PROTECTED POWER BREAKER DOCUMENTATION           **
12.        **       ---------                                       **
13.        **                                                       **
14.        ***********************************************************
15.        ***********************************************************
16.
17.
18.             1     1    2                              5
19.       1     0     6    0                              0
20.
21.        *****************************************************
22.        *                                                   *
23.        *        PANEL MC - MAIN POWER FEED PANEL           *
24.        *                                                   *
25.        *             (BREAKERED FROM BASEMENT)             *
26.        *                                                   *
27.        *****************************************************
28.
29.       BREAKER  AMPS  PH   CONNECTOR                   EQUIPMENT
30.       -------  ----  --   ---------------------       ---------------------
31.
32.       MC-1     400   3    NONE                        MG SET #1
33.
34.       MC-2     400   3                                SPARE BREAKER
35.
36.       MC-3     400   3    NONE                        PANEL MC1
37.
38.       MC-4     400   3    NONE                        PANEL MC2
39.
40.       MC-5     400   3    NONE                        PANEL MC3
41.
42.
43.
44.        *****************************************************
45.        *                                                   *
46.        *        PANEL MC1 - MAIN COMPUTER ROOM             *
47.        *                                                   *
48.        *             (BREAKERED FROM PANEL MC)             *
49.        *                                                   *
50.        *****************************************************
51.
52.       BREAKER  AMPS  PH   CONNECTOR                   EQUIPMENT
53.       -------  ----  --   ---------------------       ---------------------
54.       MC1-1    20    1    HUB. 4700 (POWER MATRIX)    PERFORM. MONITOR
55.
56.       MC1-3    20    1                                SPARE BREAKER
57.
58.       MC1-5    30    3    R&S 7324                    3350 CONTROLLER
59.
60.       MC1-9    30    3    HARDWIRED                   PDU
```

(*a*)

Fig. 8-9(*a*) Power documentation (page 1).

power grid can be easily documented. If your installation does not have a power grid and also has a relatively small equipment base, power documentation at this level might be more a burden than a help. If you do have a lot of breakers, though, this documentation can be quite useful.

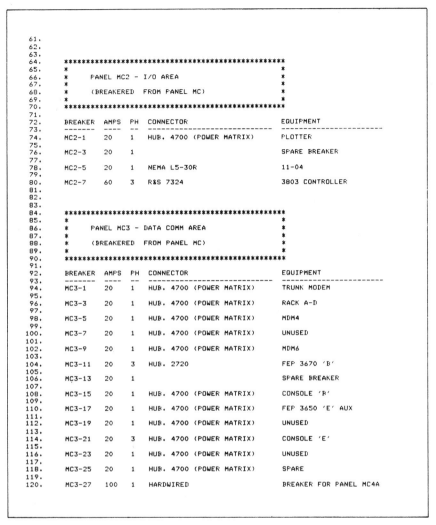

(b)

Fig. 8-9(b) Power documentation (page 2).

RELATED TOPICS

CHAPTER 9

Up to this point we have dealt with the more tangible aspects of data communications: how to design facilities, put networks into them, and work with the physical intricacies which make a data communication system work. This chapter closes the book by looking at some of the intangible aspects: the people and programs which are closely linked to the practical level. The subject would not be complete without at least a cursory glance into people and the way they go about their business. In some respects, this area may be the more important, but for the purposes of this book, we are trying to stay close to the practical level and not delve too deeply into the more philosophical things.

Therefore, the intent of this chapter is to discuss a few of the intangibles which closely influence the people working in data comm and are of immediate importance to them. The network support people who are simply trying to do their job look in two directions: one to the job they are supposed to accomplish as they see it and the other to their management, who may see things a little differently. This leads to two obvious candidates of discussion: maintenance programs and management guidelines.

The network staff may largely exist because equipment does break down and someone is needed to point the finger at the faulty component. How faults are resolved and how well they are resolved are determined by the type of maintenance strategy and program. This is a very important area of concern to those who work daily at the practical level. Inadequate or misdirected maintenance programs can be very frustrating to work under. At this level, the staff may have an opportunity to correct or offer input into the maintenance programs, since they are in such intimate contact with the subject. At the management level, though, direct involvement of the staff is less likely, and frustrations develop at times when management appears to be aloof to the problems at the staff level. We need to look into this a little, but first we will take a short look at the network support staff itself.

SUPPORT PERSONNEL

We started the book by saying that data comm is a lot of things to different people. Even after focusing on the particular type of data communications we have been discussing, there is still a large variety of skills needed to support a network. There is the software side at the host level: session control, generation of the host system for communications, and the like. Software also exists in the network, starting at the communications controller of the host and incorporating itself into network node processors—and even down to the smaller components of the network such as contention boxes and modems.

Exactly how much software knowledge is required of the network staff depends heavily upon the individual organization, its goals, and the services it offers. Close to the software is the character level information which flows over the network. You see it over the diagnostic equipment screens all the time. It is fair to say that this level of knowledge, i.e., the character level protocols and handshaking, is required of the network staff if they are to do their jobs effectively. This level is generally a starting point for the network staff to diagnose problems; and for knowledge required, it lies somewhere between the pure hardware and software areas. Decomposing the information further, the bit level brings one closer to the hardware, where the bits must be represented by voltages and currents for communication over the network.

As with the software side of the house, how much hardware knowledge an organization needs depends heavily upon its goals. An understanding of the hardware level of communications by the network staff is useful in the same sense that knowledge of higher-level software is useful in understanding the overall data communications process. It is generally rare to find the network staff knowledgeable in both hardware and software, and it is even more rare to have one individual with this knowledge. So how do you go about staffing a network? What kinds of skills are really needed, and what can you get by with?

We need a focal point to get started. Over the last several years the International Organization for Standardization (ISO) has developed a layered approach to defining some of the tasks involved in data communications and suggests that these layers be standardized to allow multivendor compatibility. This "Open Systems Interconnection"* architecture consists of the following seven layers:

       7 Application
       6 Presentation control
       5 Session control
       4 Transport (end-to-end control)
       3 Network control
       2 Link control
       1 Physical control

These levels form a hierarchy of data communication tasks starting from the very basic physical requirements (1) for talking over some media to what it is that all these levels are supposed to make possible, i.e., the application (7). The reason for showing the ISO model here is to provide a basis for the areas a typical network staff might be responsible for. That is no easy issue to resolve because of the variety of internal politics, funding problems, available staff, etc., which determine how each organization is structured.

It has been my experience that most people feel that the network staff should be responsible for levels 1 through 4; further, that includes both the hardware and software aspects of those four levels. Very generally, this encompasses the

*Consult the latest version of *Reference Model of Open Systems Interconnection*, ISO/TC97/SC16/N117.

communication controller out to the end user connection and would be the proper domain of the network staff. Undoubtedly, the majority of organizations do not fit this model, but it is perhaps a model which we can keep in mind for future changes in each of our network organizations. In a little more detail, the lower four levels are loosely defined below. Levels 5 through 7 would be generally considered host-level tasks, under the responsibility of system's programmer level personnel.

LEVEL 4: The Transport Layer

The function of the transport layer is to control transport of information from node to node across the network. It hides the fact of how it is done from the upper levels, i.e., how the topology of the network influences session level interchanges of data. The transport layer would also be concerned with optimizing the available communications options within the network in terms of performance and cost.

LEVEL 3: The Network Layer

The network layer provides the control between the network nodes established by level 4, such as between a terminal user and the access point of a public packet network.

LEVEL 2: The Link Layer

The link layer provides the protocols and functions for communication over a single data link between two systems. Controlling communication over a point-to-point or multipoint circuit would be an example. Level 3 would establish which link; level 2 would handle the details of link layer control; and level 1 (following) would handle the actual physical details for communications.

LEVEL 1: The Physical Layer

The physical layer supports the connection of a network node to the communications media, sometimes referred to as level 0. Rules for the electrical, mechanical, and signaling (handshaking) interface are defined by the physical layer. RS-232 would be an example of level 1 control between a terminal and its modem.

We now have a model of potential tasks and knowledge which might be required of the network staff, but we need more to formulate an answer to what type of staff is required to do the job. The model at this point would be useful only for defining staff requirements for the troubleshooting phase of maintenance. At what level should the staff be involved in the actual repair activities of network components and systems? Should they repair defective modems? Should they only finger-point the failing device and then call in outside maintenance people for its repair?

How such tasks are accomplished has a large bearing upon the qualifications of the network staff. At one extreme would be a staff required to perform all troubleshooting, maintenance, and other activities internally to support the four ISO levels of their network. This would be the maximum commitment for a network organization. At the other extreme is a staff which needs enough knowledge to understand the four ISO levels, but only to the degree necessary to point the finger at the failure point and then call in repair specialists for its remedy. This would be the minimum requirement for staffing the network. Decision points for selection of network staff often boils down to choosing between people with strengths in either hardware or software; seldom do they have both. By the very nature of the network, you need both disciplines covered, but for practical reasons, it may be impossible to cover them to the degree desired.

The Minimum Staffing Effort

Considering the minimum staffing effort first, what kind of individual do you need to support the network? Chances are that this organization has a small equipment base, a small budget, not so critical maintenance response time requirements, or some combination of these attributes. Since it is likely that small numbers of network people are involved, maybe even one, you may not have much choice. Look at your organization; determine how it relates to the ISO model; and then pick people with relevant skills if possible.

If you have a lot of level 0 and 1 activity, you need someone who understands a little electronics and has experience in using tools and equipment. Also, consider that your organization may already have someone knowledgeable in software but not in hardware. Choosing someone with hardware experience can round out the organization better. Consider training too. If you have hardware requirements, it's probably easier to train the hardware person to pick up the necessary minimal software skills than to train the software person to pick up the necessary minimal electronic knowledge. The programmer needs to pick up the basic knowledge to understand how a circuit really works electrically. Additionally, the experience and ability to work with tools and components would have to be developed by a software person.

An example may help to make the point. If you work a lot with 327X coax, the system itself will help to troubleshoot an ailment to a degree. 327X diagnostics will let you determine the rough location of the problem without any skills in software. But to get to the root of the problem may take more diagnostic help than the built-in 327X diagnostics can provide. Here an understanding of grounding or noise problems and mechanisms is necessary, which a typical programmer needs to go back to school to learn.

On the other hand, your network may have the level 0 and 1 details handled by others. In this case, you might want to consider network troubleshooting staff whose talents are more directed to network software. Chances here are that your staff will be responsible for resolving a network problem, whether they personally do it or not. They will need to know the software side of the business thoroughly to make an initial determination of what additional help, if any, is needed. When you rely on others to do your hardware maintenance, fast and accurate problem finger-pointing will get hardware problems solved with minimum customer inconvenience.

The Maximum Staffing Effort

At the other extreme of network staffing, where all the troubleshooting and maintenance activity is in-house, the decisions on staffing may be less troublesome. The staff must be larger to support the additional work load. This makes it possible to have in-house specialists in both hardware and software, which is a nice way to run a shop. Experts in both hardware and software make an organization very self-sufficient and able to operate in a cost-efficient manner. In a situation like this, it works well to have the mainline network troubleshooters trained at a general finger-pointing level. The experts can be called in to assist in the more difficult problems as needed. This approach optimizes the utilization of personnel and lets each one work at capacity and desired specialization.

MAINTENANCE PROGRAM

A maintenance program to cover all four ISO levels involves a lot of equipment: front ends, node processors, diagnostic equipment, contention units, multiplexers, modems, and perhaps even the end user terminal equipment. Some of the factors which go into making decisions about the maintenance program to cover this equipment are:

- Available personnel and staffing levels
- Proximity to vendor maintenance support
- Downtime constraints
- Availability of appropriate vendor maintenance programs such as per-call maintenance or ship-for-repair maintenance
- Complexity of the equipment
- Complexity of the spare equipment and components complement

It is impossible to itemize the various types of data comm organizations and suggest a proper maintenance program for each. The best we can do is discuss the basic maintenance goals every organization should consider and then conclude with general comments about vendor and in-house maintenance programs.

Basic Maintenance Program

The minimum provision for maintenance that an organization can have is a network staff which operates solely in the troubleshooting mode and turns maintenance of defective equipment over to either on-site or off-site vendor support. At most, the network staff might be responsible for some preventive maintenance (PM) activities. What would be a minimum maintenance program under these circumstances? The most obvious requirements are a good troubleshooting program, an internal structure to efficiently handle and control vendor-supplied maintenance, a viable preventive maintenance program, and the ever-present documentation effort.

It should be obvious, however, that it is impossible to operate solely in this mode of maintenance. There will always be times when an in-house maintenance effort is required. For instance, you wouldn't necessarily call in vendor maintenance to reseat a loose connector. Another example would be where a critical problem forces your staff to fix some equipment while taking instructions over the telephone from the vendor. More generally, many organizations have at least some small internal twisted pair or coaxial network which is going to require an in-house maintenance effort to support it.

These four areas for basic maintenance support will now be discussed in a little more depth. The activities listed would be appropriate to any network organization, regardless of how little or how much of the maintenance is done in-house.

1. *Troubleshooting Program*

If you cannot find the problem, repairing it is difficult. A good troubleshooting program and policy is an essential part of the maintenance program. Does your network have good diagnostic equipment and aids to help the staff? Is the staff using Telco troubleshooting services efficiently? Telco and others also offer helpful training classes for network troubleshooting. Do your equipment vendors offer troubleshooting services or training to make your own staff more proficient? Troubleshooting is similar to programming. You cannot write a cookbook on how to do it; you need a certain mentality to do the work. Your best attack is through proper training and having the right diagnostic aids available. The rest is up to the staff.

Does your staff know how far they should go in their troubleshooting efforts? Should they stop when a problem is shown to exist outside the physical boundary of your shop? Should they help customers troubleshoot a problem at their end simply because they pay for your network services? Are your customers aware of the limits and boundaries of the troubleshooting services offered with your network service? If the customer owns the equipment on your end, will you troubleshoot past it? Will you troubleshoot nonstandard equipment on the network? Do you provide troubleshooting for the initial equipment installation to get a new customer running? Will you troubleshoot a dial problem at the customer end to resolve whether the coupler or terminal is defective? Do you have internal procedures for troubleshooting when several groups within your organization are required to isolate a problem? These are a few questions which can be addressed to determine the adequacy of your troubleshooting policies.

2. Basic Vendor Maintenance

Vendor maintenance activity in a basic maintenance program would probably involve either per-call maintenance, where the vendor comes to the site as needed, or ship-for-repair, where you send the faulty item off to the vendor for repair. Vendor maintenance at a more involved level will be considered later. An obvious requirement for ship-for-repair programs are spares to replace defective units while they are being repaired. Other than that, the chief concern for implementing a basic vendor maintenance program is proper record keeping:

- Record when the malfunction occurred.
- Note the details of the failure: its frequency, symptoms, etc.
- Record when you notified the vendor for repair or shipped the defective unit for repair.
- Record the serial number of the equipment sent in for repair.
- Record when the unit was repaired on-site or when it was received from the vendor.
- Record who it is you are supposed to call for either per-call or ship-for-repair maintenance. Will you call during odd hours? If so, ask who to call under these circumstances.

By recording this information, you will have a good audit trail for the times when things go sour. You may find that the vendor cannot seem to fix the problem, and the records will show if the trouble is a repeat problem or a series of related problems. The records will also show the performance of both the equipment and the vendor-supplied maintenance for future considerations.

3. Preventive Maintenance Program

Developing a preventive maintenance program is fairly straightforward. Essentially, all you need to do is gather up the equipment manuals for the equipment you are responsible for and make a list of the preventive maintenance activities recommended by the manufacturer. In more detail, you might consider forming a PM notebook with sections covering:

- Vendor equipment PM procedures. A copy of the manufacturer's instructions would suffice.
- In-house PM procedures. This should include PM efforts for your documentation as mentioned in earlier chapters.
- A schedule on how often each of these activities should be performed.
- A checklist to show when you actually performed the PM activity.

4. Documentation Efforts

Most of the documentation efforts have already been mentioned in the preceding sections or in Chapter 8. They are mentioned again here to stress their importance. As part of the troubleshooting program, you might keep a log of troubles showing time of occurrence, a brief description of the problem, and the problem's resolution. This will help you determine repeat problems and unreliable points in the system, and it may help you resolve problems between you and your customers when a conflict over maintenance develops. It will also help show the work load placed upon the network staff and justify changes to the program such as adding new personnel.

Vendor Maintenance

The purpose of this section is to address the more involved vendor maintenance programs, principally on-site vendor support. This kind of program may be desirable for large pieces of equipment such as front ends or node processors or extremely critical equipment, or when a number of similar pieces of equipment are installed. Basically, the on-site vendor-supplied person becomes a part of your support staff, for which you pay a monthly fee for services. The amount of control you have over the person will vary among vendors. By keeping the records mentioned in the basic maintenance program, you will be in a position to judge and prove the quality of the service. The following are some of the major advantages to selecting an on-site vendor maintenance program:

- The monkey is on the vendor's back to maintain a qualified person on-site to perform maintenance. When you do it yourself, you must consider the additional staff required for vacation periods, training periods, and illness.
- Odd shift and weekend work is not a pleasure for anyone. On-site vendor maintenance helps overcome some of the problems with your own staff associated with weird working hours.
- Personnel turnover with 2 weeks' notice can disrupt your service program immensely when you do the maintenance yourself.
- Having vendor-supplied maintenance ensures that you have the full support of the vendor. If you perform maintenance yourself, you may or may not get the help you require from the vendor. If the vendor supplies it, there is no question that the vendor is responsible for resolving a problem.
- Vendor-supplied on-site maintenance avoids the hassles involved with maintaining spare parts and shipping or receiving defective or repaired parts.
- In some cases, equipment upgrades and engineering changes are more likely to be taken care of by on-site vendor maintenance. If you do the maintenance yourself, you may be lucky if the vendor will let you know of improvements. If the vendor has to do the maintenance, the company will often install an improvement to minimize its own maintenance effort.
- Vendor-supplied maintenance will probably lead to the fastest solution to prob-problems. The vendor's only purpose is to maintain your equipment, and the company's only commitment is to that equipment. In-house maintenance often

dilutes the expertise of those doing the maintenance because they are assigned to other tasks and cannot keep up their ability to solve equipment problems. There is another problem in this regard: equipment fails so seldom that it is very difficult to keep in mind all the things you need to know to solve a problem when it does occur. Vendors can cope with this more readily, since they are involved in many more instances of failure than you are at your own site.

In-House Maintenance

When your network grows and the amount of equipment starts to build up, you start to consider taking on the maintenance functions yourself. The number of people involved in-house for network support gets large enough to start countering some of the reasons for having the vendor do it: you have enough staff to take care of individual absences due to training, vacation, and illness. You probably are in a multishift operation at that point, so those related problems are already in existence. You are already handling the shipping for the smaller stuff that you support, so that is not a problem. So why not maintain the larger equipment too?

Under these circumstances there are economic incentives to do your own maintenance. When you observe what the vendor-supplied maintenance person actually does, you find that there is considerable free time because the equipment does not fail very often. Why not take advantage of it? Doing your own maintenance can allow you to utilize staff which you would have anyway and add maintenance responsibility to their duties.

An example may help. If you have vendor-supplied on-site maintenance for your front-end or node processors, chances are you can use that person only for maintenance purposes, you cannot use the person for troubleshooting the network. So you must keep troubleshooters on your staff in addition to vendor maintenance people. If you train your own people to do the maintenance, you can increase the efficiency of the maintenance program and save money. Under these circumstances, you are probably a large customer of vendor equipment, so the vendor is more willing to help you out in the maintenance activities or run the risk of losing you as a customer.

Backup is a key point in performing the maintenance functions in-house. You must have a backup structure to cover the times when the primary maintenance person is unavailable. Third-level backup may be advisable under certain circumstances, such as when you maintain several pieces of similar equipment and the primary and secondary support people are occasionally swamped with work. It is a good idea to publish a maintenance call-out list for interested parties within your organization so that it is clear who should be called to maintain each item of equipment. This will also make it clear to the support staff who is responsible for what.

The list should show each piece of equipment, who is the principal person to call for maintenance, and who are the backups in case that person cannot be reached. For critical equipment it is a good idea to have the last backup knowledgeable enough to call the vendor for assistance. Try to establish an agreement with the vendor whereby you can call occasionally for assistance even though you do your own maintenance. Of course, it is only fair that you should be willing to pay for this assistance. You won't need it often, but when your primary maintenance people are unable to fix a problem, vendor backup at the last level is a valuable asset.

Doing your own maintenance also means you need to maintain spares, either board level, component level, or both. Board-level spares are the most workable; shooting problems down to the component requires sophisticated test equipment and good technicians or engineers. Only very large organizations can afford to do component-level maintenance effectively. In any event, the problem you must address with spares is that there is no point in keeping them around if they don't work. That means that you must maintain a program of rotating spares into the operational equipment to ensure that spares on the shelf work. What have you gained by stocking a large supply of spares, some of which may sit on the shelf for literally years, and then having a spare inoperable when it is really needed?

Another problem with spares is revision levels. Don't forget to upgrade the spares along with the operational equipment. Rotating your spares periodically will occasionally turn up a board that is not up to the latest revision level. For your rotation program, it is a good idea to swap out only a few boards at a time and mark the boards that you do swap. You run a very good chance of causing your operational equipment to fail when you rotate spares into it. When this happens, your first clue is that you just recently rotated spares, so the problem becomes one of locating the failing board in the lot you just rotated. Marking the parts rotated makes it easier to identify the rough locations of the failure.

Lastly, for the spares rotation, determine the best time to do the swap. For instance, you may have light network activity over the weekend. This, then, would be a more preferable time to have a failure. In this case, rotating spares on Friday nights will give you a couple days for the parts swapped to show malfunctions. If they do malfunction, you can perform the maintenance during the lightest activity period.

MANAGEMENT ISSUES

Any discussion about management is starting to get away from the practical issues which are the theme of the text, but as mentioned earlier, it is a subject close to the everyday concerns of those who work at the practical level. Like it or not, the job we all perform must meet management's approval. Offhand, I can think of three reasons for including a brief discussion about management:

1. An understanding of some of the concerns of management may help those at the support level better understand why some decisions are made the way they

are. Your management may be remote, causing the decisions coming down to lack immediate explanation of the reasons behind them.
2. Many of the people in network support will eventually go into some type of management. A discussion on the management side will be of interest to these people.
3. Many of the people now in direct management of network operations have risen through the ranks via the software route. They have never had an opportunity to delve into many of the concerns we have been discussing in this text, but they are in the position of making decisions for the network. There can be considerable friction between the support staff and management in this situation. A few guidelines may be of use to these managers to help them understand the concerns and frustrations of the support staff who must maintain the network under their directives.

What Is Expected?

Does the network support staff know what is expected of it? This can come only from management. What is the quality of work expected? What is the quantity of work expected? What are the knowledge areas expected for the various positions in the network support staff? Are any special working relationships expected, such as between the staff and customers? For organizations which have periodic personnel performance evaluations, these areas can be addressed there and made clear to the staff. If they are not made clear somewhere, how is the staff supposed to know what is expected? They cannot read minds any better than management; these expectations should be spelled out clearly.

Neat and accurate work should be expected of everyone. Data communications is a craft with a lot of detail. If you don't pay attention to the detail, your efficiency and system reliability will suffer. Poor craftsmanship in physical assembly details, such as building cables with connectors, may not be noticed immediately. After a while, though, intermittents will begin to show up in solder or crimp joints and will haunt you for a long time. Sloppy documentation is another area which causes many problems. You simply have to keep on top of these details, or eventually you will pay a painful price to remedy the problems the lack of attention caused.

Responsibility must also be clearly placed. The support staff was hired to be responsible for something. Is it clear what that responsibility is? The call-out list mentioned earlier is a good place to put in writing what duties certain individuals are responsible for. By listing the various maintenance and troubleshooting categories with corresponding people to call for help, you make it reasonably clear where the responsibility lies. Even then, however, responsibility must sometimes be spelled out in more detail. The call-out list may not show in enough detail whom to call for a problem; the problem may not be well enough defined to place blame on a certain piece of equipment, for instance.

In a situation like this, the people using the call-out list obviously need help to solve a problem, so they could very easily call someone the problem really does not fit. The person called to remedy the problem, after digging into it, may determine that the problem is someone else's and walk off, leaving the person who called him holding the bag. It should have been clear to the maintenance person that responsibility is not so sharply defined as to allow that kind of action; there was an inherent responsibility to help determine who else should be called.

Responsibility means seeing to it that the job gets done or put into the proper hands, regardless of whether you personally do it or not. If a job is not yours, it is still up to you to see that it gets turned over to someone who can take responsibility for its completion. This problem is evident when the staff begins to play the finger-pointing game. That is an unacceptable practice to allow in your own shop; it's bad enough to have to contend with it among vendors and other organizations. When your whole staff chips in to see to it that a job gets done, finger pointing does not exist.

Decisions

I have had a quote lying around my office for years which is quite appropriate for this topic. Unfortunately, I no longer know the source. My apologies to its author:

1. Managers make decisions.
2. Any decision is better than no decision.
3. A decision is judged by the conviction with which it is uttered.
4. Technical analyses have no value above the mid-management level.
5. Decisions are justified by benefits to the organization; decisions are made by considering benefits to the decision makers.

Managers from the network level on up need to take points 1 and 2 seriously. Few things are more frustrating than having decisions hang for no apparent reason. If they must hang, it should be made clear to the staff why they are being delayed. The staff is always anxious to have decisions made promptly on issues of immediate concern to them. If they understand the delay, they will most often support the lack of a decision. If no reason is given for the delay, the staff becomes frustrated working around problems which a decision may solve.

Some decisions are difficult to make. If things come to an impasse, a straw vote among the support staff may help. If the decision turns out bad, the staff will then be more supportive than they would be if they had not been party to it. Obviously, there are other times when a decision cannot easily involve a vote among the staff. But whatever—make a decision!

The fourth point above sometimes takes a while to dawn on some of us. A network manager is a technical manager. It does little good for this person to provide technical material to upper management for use as decision points. Technical information must be buried or disguised, and the line of reasoning for a recommended decision must be couched in terms upper management will understand—in economic impacts, for instance. This can be one of the unpleasant areas where the network manager must change hats and use political or other persuasion rather than pure technical expertise. The staff may also have a difficult time with this; they are used to talking in technical terms all the time with their network manager and cannot understand why that line of reasoning cannot be carried higher by their boss. As an example, it may be technically very clear that to change to programmed jacks rather than keep on using permissive jacks for dial data sets could offer one or two dB more signal and increase the integrity of the signal correspondingly. It may cost only a few hundred dollars for the change. Management sees it differently: why should money be spent for an improvement which hasn't been needed for years?

Decisions should not be made in a void. You must have adequate information to make a proper decision. At times, that information is difficult to come by. This leads us into the controversial subject of consultants. When you hire a consultant, it has at least two immediate impacts:

1. You might look like a fool. If you were up on your job, you would not need a consultant. (This is a typical response.)
2. You have no confidence in your staff to do the job; that's why you hired the consultant. (Another typical response, often from the staff.)

The only real point to be made here is that one person cannot be an expert in all the areas of data communications. Not even your combined staff has the necessary knowledge and expertise to decide some issues. You and your staff live in somewhat of an isolated world: your own organization. A consultant lives in everyone's world and sees many different solutions to problems. That provides a degree of foresight that may be missing in your own shop. You can spend a week or two wrestling with a decision on a major program, make the decision, and probably be right most of the time. But chances are it is a shallow decision. It may not last long before technology outpaces it.

That same amount of effort could have paid for a consultant in many cases. Sometimes all you need is a few hours of talking over a problem to arrive at a clear path to a decision. The consultant can keep you on the more correct trail and keep you from wasting time on immaterial things. When you could use some outside technical assistance, don't be afraid to suggest it. You are a bigger fool for making a bad decision which could cost thousands, whereas a few hours with a consultant could have cost only hundreds.

A couple more points should be raised before we close this discussion. Sometimes a decision hangs on balance and needs just a little something to swing it one way or the other. Growth can be a decision point; always opt for a decision which allows growth. If brand X multiplexer allows future expansion compared to brand

Y for about the same price and everything else is about equal, choose *X*; in the future, you might be glad you did.

Another common decision point is selecting methods of implementation. Opt for the most standard and flexible way, even if it costs a little more. You can implement a polled multipoint circuit locally by literally hardwiring all the lines together. It will even work, for a while. But as the number of drops change, a slight change in the error performance may become visible until, at some point, the whole circuit fails. What are you going to do then, with all those people screaming to have their problems solved? Implementing it correctly in the first place, with bridging transformers, would have prevented that problem. Don't play implementation games on a production network! Here's another example. If you have a large equipment rack only partially filled, don't clutter it up with extraneous stuff so that you can't use it in the future for additional equipment.

Policies

The network manager, or higher, must make policies sooner or later. If policy is not set, the people in the organization will look very unprofessional. They will flounder around trying to decide what they think they are supposed to do; they will offer conflicting information both internally and to the customers. They may become gun-shy of working under no policies after getting burned a few times because of improper decisions they made without a policy to guide them. Chapter 8 offered several areas within data comm where policies are good to adopt.

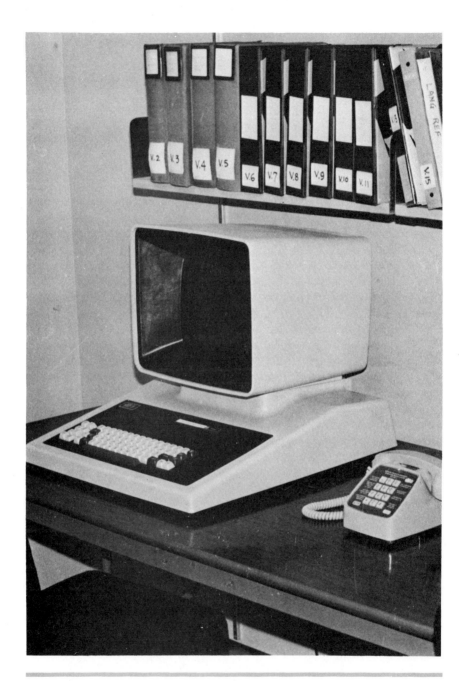

APPENDIX

The Appendix covers material in two principal areas:

1. A checklist section
2. Project implementation aids

The checklists were formed by condensing the material in the preceding chapters; they are intended to spare you the effort of scanning the entire text to check items of concern which may develop in your own projects. They can also be used as an aid in planning tasks to consider in future projects. The implementation aids will be useful in carrying out some of the ideas described in the text. The intent is not to provide an exhaustive source of part numbers and vendors; rather, it is to concentrate on the more hard to find items. I have included vendors with whom I have had experience to give you a lead to things you may need. Second sources are not necessarily given, which does not imply that the vendors listed are the only ones available or the most capable ones around. In most cases, identical or functionally equivalent items can be purchased from several reliable manufacturers.

In keeping with the adage that a picture is worth a thousand words, I have included the numbers of relevant figures for the projects under consideration. In some sections, a summary of all figures which may be of interest to the topic at hand is given. This is then further broken down to more specific figures as necessary.

NETWORKING CHECKLIST

Telco Considerations

1. If at all possible, avoid mixing telephone and data circuits.
2. If Telco shares your twisted pair wiring panel, use unique patch wire colors to show ownership and responsibility.
3. Try to talk Telco into using better-quality transient suppressors (e.g., three-element gas discharge tubes) on your data circuits rather than their customary carbon block suppressors.
4. If you still use FDM circuits, how do you terminate them? Do you bridge all but one and use the last one to terminate the circuit? Doing so causes problems when you need to swap the terminating modem. If you set all modems to the bridging state and terminate the line resistively, it will solve this problem. You can also strap your spare modem the same as those in use in the circuit; then there are no special cases to mess you up. All you have to do is install fixed 600-ohm (or whatever is appropriate) resistors on the circuit pairs.

General Local Networking (In-House)

1. Don't lay cable where people will be walking on it.
2. Don't mix power and data cables in the same conduit.
3. Don't parallel data and power runs for long distances.
4. Avoid laying cable near equipment which draws large amounts of electric current.
5. Fill conduits as full as possible in one pull.
6. Use a large pullwire for plastic conduit.
7. Use cable grease or talc, if necessary, to make the cable pull easier and reduce the stress on the cable.
8. If you are using PVC conduit, pull cable slowly to reduce heat buildup from friction. This is especially true if you use power pullers.
9. Add a service loop to all your cable runs.
10. Add a service loop when attaching cable to movable furniture.
11. Add a drip loop in areas where there is a chance of water getting on the cable connectors.
12. Clearly label facility termination boards as to ownership, especially in such areas as facility service tunnels.
13. Label all feeders, terminal blocks, and circuit drops.
14. Use strain reliefs on vertical cable runs.
15. Use feeder cable hangers which have a wide area upon which to rest the cable.

Local Twisted Pair Networking

1. Standardize on the use of 24 AWG wire size as much as possible. Exceptions would be for terminal drops and long interfacility runs, where 22 AWG can be used.
2. Use cable which has the same color coding as standard Telco cable.
3. Don't punch stranded wire onto 66B3 blocks.
4. When your wire patching panel is shared with others, use a unique wire color for each user of the panel system to show ownership and responsibility.
5. Maintain twisted pair throughout the circuit. For instance, don't use non-twisted for the terminal drop.
6. Adopt the philosophy that every circuit must link to the central wiring panel. No exceptions.

7. Install transient suppression on all interfacility circuits.
8. On long cable runs, consider using several smaller feeders rather than one large one to reduce signal crosstalk.
9. When running circuit cable, it might be wise to check if the insulation type is allowable. Some codes may give you trouble with PVC insulation in certain circumstances. In a fire, PVC can emit harmful chlorine compounds.
10. In some situations, code may require flame-resistant blocks. They are available, and they are called self-extinguishing blocks.

Local Coaxial Networking

1. Always use coaxial connectors for splices; don't try a solder splice.
2. Insulate all 327X cable splices.
3. Form coax feeders in multiples of eight.
4. When making coax feeders, leave cable ties loose enough to allow the individual cables in the bundle to slip when turns are made.
5. Consider using patch panels for terminating the coaxial data circuits.

Terminal Drops

1. Use stranded cable for terminal drops.
2. Don't custom-cut the drop cable length. Leave extra length for future changes.
3. Use a strain relief on terminal drops.
4. For no-modem networks, don't modify the terminals to fake out the EIA handshake. Do it in the terminal drop cable.
5. When a terminal is removed and the circuit stays, document it; don't lose track of it.
6. Use standard-length extenders, don't make a bunch of oddball lengths.
7. Make extender cables with opposing genders on the ends.
8. Be safety conscious; don't leave a lot of cable lying around where people can become entangled in it.

EIA Cabling

1. Think ahead how to dress the cable run and where to stash the extra length.
2. Don't use short EIA cables; the standard 50-ft (15-m) length is a good one to use.
3. Consider a cable hanger system to solve your EIA cable mess.

Equipment

1. Set a policy of strapping your spare equipment in the same way as the production equipment. When you are troubleshooting a problem is not the time to worry about straps.
2. If you use shelving similar to that described in Chapter 5, install the labels on the shelves, not the equipment.
3. If the equipment power cord plugs in under the raised floor, tie the cable to the conduit to prevent it from being disconnected inadvertently.
4. Don't bury your equipment to get it out of sight. The result may look nice but it makes working with the equipment difficult, may cause heating problems, and is often an inefficient use of space.

FACILITY CHECKLIST

Terminal Areas

1. In data-intensive areas, provide cable access to all rooms in that area. Don't leave a few out just because, in your present plans, those rooms aren't used for data.
2. In the office, try to install the power, data, and phone service triad on two opposing walls at a minimum.
3. Have data and telephone conduits lead from the room to the hallway just outside the room.
4. Don't mix telephone wiring and data wiring in the same conduit.
5. In large terminal areas, do the equipment layout with a minimum of two egress paths for safety. Check with local safety people on this.
6. Consider installing spare conduit back to the power panel in terminal labs just in case you may need special power in the future.
7. Leave an area in terminal labs accessible to handicapped people. Close to the entryway with room to maneuver a wheelchair would be a good design goal.
8. Consider left-handed people too! The work space needs to be on the side of the terminal opposite that for right-handed people. In a large terminal lab, it is easy to have one or two tables implemented with the terminals shifted from their usual right-handed position.
9. Also in large terminal areas, don't have cable lying around where people can become entangled. This is a safety problem which should not exist.

Cable Paths

1. Several smaller conduits are more practical to install than one large one.
2. Slope all interfacility conduits to drain water.
3. Make gradual bends in conduit if you have to make any bends at all.
4. Label all conduits as to length for future reference.
5. In plastic conduit, use a large nonmetallic pullwire such as hemp or plastic marine rope.
6. Install plastic bushings on conduits to prevent cable damage.
7. Try to have cable trays on both sides of hallways.
8. Install occasional crossover trays from one side of the hallway to the other.
9. Telephone and data wiring can share the hallway trays to help justify the cost of the trays.
10. Don't forget space for vertical cable runs.
11. Consider ways to strain-relieve vertical cable runs; install a board or something similar for cable clamps.
12. Consider ways to eliminate sharp corners where vertical cabling will hang.

Comm Area

1. Design and lay out the comm area with cable flow in mind.
2. The equipment in the comm area should be seriously considered for protected power, such as a UPS.
3. Consider using a power grid under the raised-floor area of the comm area.
4. The comm area should have its own breaker panel.
5. Consider using flex conduit for special power runs. If you use it, don't custom-cut it; leave extra length for future changes.
6. One breaker per receptacle increases reliability.
7. Use of twist-lock power connectors decreases the chance the power cord will be accidentally unplugged.
8. Tie power cords down.
9. A good depth for raised floor is 12 to 18 in (30 to 50 cm).
10. A good tile size is 2 ft (60 cm) square.
11. Consider standardizing the way you cut floor tiles.
12. Keep water pipes out of the raised-floor area.
13. Seal concrete floors to reduce dust.
14. Don't use ionization detectors under raised floors.

15. Consider using a cable hanger system.
16. Consider using the central wiring panel concept.
17. Consider customizing your equipment shelves to increase their practicality and efficiency in working with them.

Cable Termination Areas

1. Design the network with service zones, each zone having a terminal block to terminate the feeder cabling.
2. In service tunnel areas, watch out for water.
3. Is room adequate for all the cable?
4. Leave space for feeder service loops.
5. In locating the termination board, consider who will be working in the area and what they might be doing. Pick an out-of-the-way place for the board if possible.
6. Label the termination area so that ownership and responsibility are clearly established.
7. A good way to store cable in the form of a service loop is to install ordinary garden hose hangers in the cable termination areas.

General Electrical Practices

1. Use stranded wire in flexible conduit.
2. Generally speaking, all electrical cable must be installed in metal conduit. Don't lay it in your cable trays.
3. Avoid paralleling power and data cabling.
4. Label breaker panels accurately and in a way to rapidly identify what is connected to them.
5. Label all power receptacles with their corresponding power panels and breaker numbers.
6. If you elect to use twist-lock hardware, you will need to make some power extension cords. Use standard lengths.
7. Have you ever had the wire screws in your circuit breaker panels checked? You might be amazed what you discover by retorquing these screws. Retorquing might even clear up some very subtle problems you have been having. It does not have to be done very often, but it should especially be done 6 months to a year after a new power installation.
8. Another thing you can do in this regard is to have your facility electrician simply go through your power feeder wiring and feel for warm spots. They can be a sign of bad connections.

DOCUMENTATION CHECKLIST

1. Overdoing documentation can be as bad as having no documentation at all.
2. Making agreements accessible to the network staff may avoid some prob-problems.
3. Consider an archive for your data comm documentation.
4. Consider a TP change request form or its equivalent to direct information to those who need it and enhance intergroup communication.
5. Consider a work-in-progress form or its equivalent to keep tabs on the work.
6. Document your bulk feeder wiring.
7. Consider blueprints for parts of your documentation effort.
8. Is documentation upkeep a line item on your PM procedure?
9. Consider putting schematics and maintenance manuals on your equipment orders as a line item.
10. Document circuits that are unused as well as those that are in use.

CUSTOM PROJECTS—CONSTRUCTION NOTES

If when reading this section you are serious about building a project, you will have to digest the related information in the text also. Here I have only a few notes of points which may not be too obvious.

1. *Adapter Assemblies*

    (Figures 6-15 to 6-18)

    Construction of adapter assemblies, as shown in Figures 6-15 to 6-18 should be done by someone with soldering experience. The other construction difficulty is the dust cover, should you decide to use one. Use a rounded dowel of appropriate size to form the Plexiglas [0.875 by 4.5 in (2.22 by 11.43 cm)] when you heat it with a hot-air gun. The ends opposite the rounded end can be bent inward slightly to hold the dust cover on the adapter when it is installed. Expect to practice a little before you produce a nice looking adapter. The screwlock/screw retainer assemblies and connectors are listed under data connectors in the next section, where you can find part numbers. Other parts are the following:

    a. Inductor (Figure 6-18):

    This is a small signal inductor (choke) which you should be able to acquire from your local electronics parts distributor. Try Allied (P/N 871-3310, 1981 catalog) if you are unsuccessful elsewhere.

    b. TCW [Figure 6-15(*a*)]:

    Use Belden 8019, or its equivalent, tinned copper wire (TCW), which is available from most electronics distributors. Cut these wires with a sharp diagonal cutter to maintain length accuracy. The length of these wires sets the width of the assembly for the Plexiglas.

2. *Cable Assemblies*

[Figures 3-12(b), 4-8, 4-10, 4-17, 6-7 to 6-14]

If you use the crimp-type parts shown in the illustrations, there should be little difficulty in making your own cable assemblies. You do need the proper tools, however, to do the job correctly. The tools are a minimal investment, so don't try to crimp parts with a pair of pliers. Since you will be working almost exclusively with stranded wire, the crimp method is much easier than trying to use solder parts. All the parts are standard and are listed in the next section under data connectors. Notice the strain relief detail shown in Figures 4-8 and 6-7.

3. *Cable Hanger Grid Components*

(Figures 5-4, 5-5, 5-6, A-1)

The cable hanger rings, shown in Figure A-1, can probably be fabricated by a local metalworker. Many materials can be used; cold-rolled rod was used here. The mounting rail is commonly available shelf standard material used for putting shelves on walls. It can be cut to length as needed. The clamps—clamps—hose clamps—are also commonly available; try Ideal Corp. through your local hardware store.

4. *Equipment Shelves*

(Figures 5-9, 5-10, 5-11, A-2)

You should turn the job of building the equipment shelves over to a local cabinetmaker. Birch ply makes a nice looking unit (try select red birch). Set the width of the shelves with floor tiles in mind. If both sides of the shelf coincide with floor tile edges, you may have a problem lifting floor tiles adjacent to the shelf. Other dimensions can be varied to suit your needs. Make

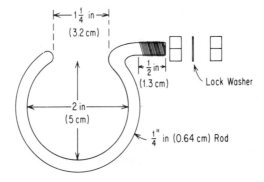

Fig. A-1 Cable hanger detail.

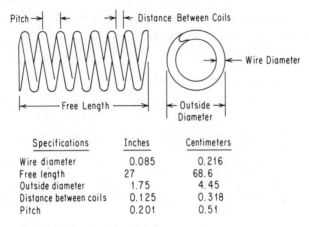

Specifications	Inches	Centimeters
Wire diameter	0.085	0.216
Free length	27	68.6
Outside diameter	1.75	4.45
Distance between coils	0.125	0.318
Pitch	0.201	0.51

Fig. A-2 Cable spring detail.

a slot in the top back section just in case you need to run a data cable into the top storage area.

a. Springs (Figure A-2):

The springs can probably be built by a local firm dealing in springs. Otherwise, try McMaster-Carr. The specifications listed work quite well for most cables found in data comm applications. A piece of conduit is installed inside the spring and screwed to the back edge of the shelf with a spacer. This allows the spring to move and also provides a rounded surface for the cable to go over. Chrome plating makes the springs more attractive.

b. KV Track [Figures 5-9, 5-11(*b*)]:

The cabinetmaker can supply this item. The tracks should be recessed into the wood so the shelves meet with the sides.

c. Label Holders (Figure 5-9):

Try Wright Line Inc. P/N 30613 works for 0.75-in (2-cm) shelves.

d. Power Strips (Figure 5-11):

Wiremold works well and should be locally available from electrical supply houses.

5. *Terminal Block Panels*

(Figures 2-6, 4-4, 4-5, 4-16, A-3)

The terminal block panels should give you no problem in building them; the figures provide most of the necessary detail. The solid-wire version uses the same punch block as on the large panel wiring system.

a. Dakota Clamps (Figures 2-6, 4-4, 4-5):

These are the square-looking clamps for holding the cable bundles. Try Dakota Engineering (P/N 2CI-50-9) for them if you can't find other local sources. Expect a 50-piece minimum from Dakota.

b. 66B3 Blocks [Figures 4-4(*a*), A-3(*a*)]:

Try North Supply. If you need the self-extinguishing type for fire code reasons, add an SE suffix. Siemon makes good-quality blocks and also has a large variety for various applications. Most are available through North Supply.

c. AMP Blocks [Figures 2-6, 4-4(*b*), 4-5, 4-16, A-3(*b*)]:

You will need both the small blocks and block base as shown in Figure A-3(*b*). It takes 2 bases and 10 blocks to build a 50-pair block panel. You can get different styles of the blocks, which may be of interest for other applications.

6. *Twisted Pair Panel Wiring System*

(Figures 2-4, 2-5, 3-17, 5-7, 5-8, 7-1)

These panels are not particularly difficult to make, but they are quite labor-intensive. The idea is to size them in such a way that a full sheet of plywood can be used for each vertical section. Not shown in the pictures are the mounting rails used to hold the panel system away from the wall so that cabling can be put behind the panel. The rails can be sized to bring the panel out from the wall the desired distance. There is not much force on the wall to hold the panel system, but there is a lot of weight put on the floor. As shown here, 45-degree holes were drilled for the 25-pair cables which attach to each block. This takes some of the bending strain off the cable as it travels through the panel from the back to the front. You will notice letters and numbers at the top and sides of the panels. This is a reference system to designate block locations. The connector ends of the cables should be labeled with their respective block number identifiers. Suffix the label with an L or R depending upon which side of the block it goes to.

a. Wire Loops [Figures 5-7(*b*), 7-3]:

These are sometimes also called drip loops. They are made by Diamond, who may refer you to a local distributor. They come in several sizes, but part number 15-00121 or 15-00122 is best suited for this application.

b. Wire Guides [Figure 5-7(*a*)]:

These can be difficult to obtain unless you build them yourself. The ones shown in the pictures are mass-produced for telephone company applications. It's hard for the little guy to go out and buy them because of the small quantity. Try talking to your local Telco data people and see if an arrangement can be made to get some from them. They are also known as D rings in the trade. You can try Comm-Fab, but expect quantity problems.

c. Cable Assemblies [Figure 5-7(*b*)]:

Work can be sped up considerably by purchasing cable assemblies with a female connector installed at one end. These are known as service extensions. They are manufactured in such large quantities that prices are low enough that it really doesn't make sense for you to assemble them yourself.

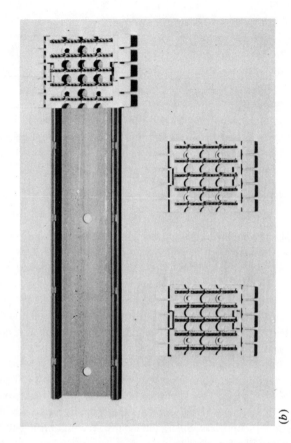

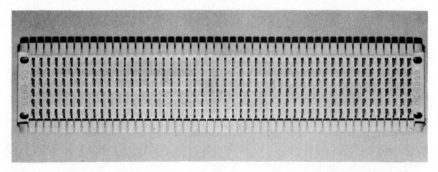

Fig. A-3 Punch blocks. (*a*) Solid wire punch block (P/N 66B3-50); (*b*) AMP stranded (and solid) wire punch block (block P/N 552499-5, base P/N 229952-1).

The cable connector end should start about 4 ft (1.2 m) or so below the panel system. You can also specify that the opposite end be stripped of the cable jacket, which will save you some additional work. Have about 12 in (30 cm) stripped if you bring the cable to the center of the block as shown here. North Supply, TW Comcorp, and Vari-Tronics are good sources for these cables.

7. *Coaxial Data Circuit Patch Panel*

(Figures 2-8, 6-1, 6-6)

You can build these patch panels in a variety of styles depending upon your need; there is nothing magic about those shown here. If you want to farm out your panel work, California Chassis does good work and doesn't seem to mind small-quantity jobs. They can do both the drilling and the painting. Don't forget to mount the coax panel connectors on an insulator; those shown here are mounted on G-10 glass epoxy unclad circuit board. Try The Mica Corporation if you don't have a local source for G-10 board.

a. Coax Bulkhead Connectors [Figure 6-6(*a*)]:
Use UG-492A/U connectors available through your local electronics parts distributor.

8. *Local Data Circuit (LDC) Panel*

(Figures 6-1, 6-2, 6-3, 6-4)

The LDC panels have a lot more metalwork involved than the CDC panels; you may want to have them made outside your own shop. Again, Cal Chassis is a source if you cannot locate someone to do it locally. The two rows of integrated-circuit sockets are mounted on a circuit board which was custom-etched for this application. The figures should show most of the other details you will need to make your own panels.

PARTS IDENTIFICATION AND VENDOR REFERENCES

1. *General Electronic Supplies*

For the common things shown, such as IC sockets, cable clamps, cable ties, grommets, terminal strips, and heat shrink tubing, take a walk through the yellow pages for local electronics distributors. You should have little difficulty in locating these items.

2. *General Telecommunication Supplies*

If you need a source for a variety of items related to the telephone industry and data communications, try North Supply or TW Comcorp.

3. Labeling Materials

(Figures 2-5, 3-17, 4-4, 4-15, 4-16, 4-17, 5-2, 5-8, 6-2, 7-3, 7-4)

a. Cable Labels (Figures 4-15, 4-17):
Terminal drop labels	Brady SLSH 1015
Small feeder labels	Brady SLSH 10375
Large feeder labels	Brady SLSH 20375

b. Block Tab Labels (Figures 4-4, 4-16, 5-8, 7-4):
Try Brady B-500 series labels. They are about the right size and are fairly easy to work with.

c. General-Purpose Labels (Figures 5-2, 6-2):
Avery makes a variety of small labels useful for some of the labeling jobs shown here. They are found in most office supply stores.

d. Lettering (Figures 2-5, 3-17, 4-16, 5-8, 7-3):
There are a few different sources for this type of lettering. The peel-off vinyl type works well. Try E-Z Letter Quik Stik. You should be able to get it locally.

4. Data Cable

(Figures 2-1, 4-2)

a. Coaxial Cable (Figure 2-1):
Use RG 62A/U for 327X applications available from your local suppliers. There are several variations of this cable, some of which are unsuitable for 327X applications. Some trouble might be avoided by specifying cable equivalent to what IBM recommends. (Indoor, IBM 323921; outdoor, IBM 5252750)

b. Twisted Pair Feeder Cable (Figure 2-1):
Try North Supply, TW Comcorp, or Vari-Tronics.

c. Twisted Pair Terminal Drop Cable (Figure 2-1):
Table A-1 lists a nice family of cable useful for this application.

TABLE A-1 Paired, Stranded 22 AWG Cable

Pairs	Alpha no.	Belden no.
1	1316	8442
2	1317	9744
3	1318	9745
4	1319	9746
5	1320	8747
6	1322	8748

d. Patch Wire (Figure 2-1):
You need this for wire patching on the panel wiring system. Use unjacketed 24 gauge twisted pair wire available from North Supply, TW

Comcorp, or similar suppliers. If you mix Telco and your own circuits on the same panel system, use different colors for the two users. Ask for jumper wire.

e. Feeder/Subfeeder Cable Assemblies (Figure 4-2):
The large feeder cable such as 200-pair should be bought with no connectors installed. Pulling this cable with connectors attached is too diffi-difficult. After pulling, you can install 50-pin connectors with a special tool available from AMP. (Connector and tool part numbers are given later in their respective sections.) You will need the cable boots to cover the wire bundles broken out into the individual connectors. The boots can be eliminated, but the end product doesn't look very good that way. Look in the parts section later on for part numbers. The smaller subfeeders are a different matter. These runs from the terminal block to the main feeder can be purchased with connectors installed on one end at very reasonable prices. For terminal drop subfeeders you need to get 50-pair cable with two male connectors installed. These are known as service feeders in the telephone trade. All this cable can be purchased from North Supply, TW Comcorp, or similar distributors.

f. Cable Assembly Cable (Figures 6-7 to 6-14):
Some of the cable assemblies have the wiring jumbled so much that you cannot use paired cable to build the cable assembly. In these cases, try the family of cable shown in Table A-2.

TABLE A-2 Unpaired, Stranded, 22 AWG Cable

Conductors	Alpha no.	Belden no.
2	1172	- -
3	1173	8443
4	1174	8444
5	1175	8445
6	1176	- -
7	1177	9430
8	1178	9421
9	1179	9423
10	1180	8456

5. *Data Connectors*
(Figures 4-1, 4-2, 4-3, 4-17, 6-7, 6-15)

a. 50-Pair Feeder Connectors (Figure 4-2):
AMP has connector kits which work well for terminating cable. You do need the accompanying tool to use these connectors, however. As the tool is rather expensive ($400 range), you might have second thoughts. I highly recommend the tool if you will be working with cable on a steady basis. It's called a butterfly tool in lay parlance; see the tool section. AMP plug

connector kits are part number 229912-1 and receptacle kits are part number 229913-1.

b. Terminal Drop Connectors (Figures 4-17, 6-7):

Here are all the little goodies you will need to make connector assemblies:

Part	AMP part no.
Plug	205208-1
Receptacle	205207-1
Shield	205718-1
Screwlock assembly	205817-1
Screw retainer	205980-1
Sockets	66505-9
Pins	66507-9

You use the plug, screw retainer, and pins to make a male connector and the receptacle, screwlock assembly, and sockets to make the female con-connector. The shield covers a connector of either gender.

c. Coaxial Connectors (Figures 4-1, 4-3):

Try Amphenol 31-321. Cables are linked together by using a barrel-type connector, part number 31-219, also from Amphenol.

d. 25-Pin Solder Connectors (Figure 6-15):

Try DB-25P (plug) and DB-25S (socket) D Cinch connectors. They are made by a variety of manufacturers and are available from your local electronics distributors. AMP screwlock assemblies and screw retainers fit these D connectors also, or you can use those supplied by the connector manufacturer.

6. *Electric Power Components*

(Figures 5-1, 5-3, 5-11)

All the following electric power components should be available locally from electrical supply houses.

a. Twist-Lock Receptacles (Figure 5-1):

Try the following Hubbell parts:

Duplex	4700
Single	4710
Cord	4729C

The cord receptacle is useful for making power extension cords. If you adopt using twist-lock components, you will have to make your own extenders and will need a supply of cord receptacles.
   b. Twist-Lock Plugs (Figure 5-1):
   Try Hubbell 4720C.
   c. Flex Conduit (Figure 5-3)
   d. Power Strip (Figure 5-11):
   Try Wiremold from your local electrical distributor.

7. *Miscellaneous Parts and Materials*
   a. Pullwire:
   For pulling wire in metal conduit, aircraft cable 0.125 in (0.3 cm) or so in diameter works well as pullwire. It should not be used in plastic conduit, however. Instead try 0.5-in (1.3-cm) or larger polyethylene rope available at your local hardware store.
   b. Cable Grease:
   Try Ideal 31-351 yellow pull grease from a local electrical distributor. Talc also works in some situations.
   c. Coax Insulation (Figures 4-1, 4-3):
   Try Tygon or similar plastic tubing, also available at your local hardware store.
   d. Transient Suppressors (Figure 6-19):
   Try GE V47ZA1 suppressors from your local electronics supply house.
   e. Plastic Raceway [Figures 3-6(*a*), 4-11, 4-13, 5-12(*b*), 6-1, 6-3]:
   A variety of sizes are available from Panduit, which sells through local distributors. Call Panduit to find the nearest supplier.
   f. Floor Cable Guard (Figure 4-12):
   Try Inmac as a source for this material. Some types have a slit cut lengthwise to accommodate cable which has connectors. This can be a handy feature for certain applications.
   g. Cable Connector Boots (Figure 4-2):
   Try North Supply for these items. They are available in various types for different sizes of feeders.
   h. TV Outlet Plates (Figure 3-1):
   Try locally available Sierra S-12. They are also called phone outlet plates.
   i. Outlet Plate Connectors:
   Computer Cable & Products makes a variety of plates with connectors installed.
   j. 66B3 Blocks [Figures 4-4(*a*), 5-8]:
   Try Siemon 66B3-50 blocks available from North Supply.

190 APPENDIX

    k. 232/449 Adapter (Figure A-4):
    The one shown is an ITT KS21253-L7.

8. *Facility Items*

    a. Vertical Access Doors (Figures 3-13, 3-14):
    Inryco, Inc. is a manufacturer.

    b. Cable Tray [Figures 3-11, 3-18, 4-1(*a*)]:
    Try a local distributor for these; Husky/Burndy is a manufacturer.

    c. Conduit Bushings (Figures 3-2, 3-10, 3-15, 4-18):
    Your local electrical distributor has these in stock.

9. *Tools*

    a. Connector Tools (Figure A-5):

- Figure A-5(*a*) shows pin insertion and extraction tools useful in working with AMP 50-pin connectors.
- The "butterfly" tool shown in Figure A-5(*b*) is used to connect a 25-pair cable to a 50-pin connector.
- The EIA tool shown in Figure A-5(*c*) is used for inserting and extracting pins and sockets in EIA connectors.

Fig. A-4   232/449 adapter.

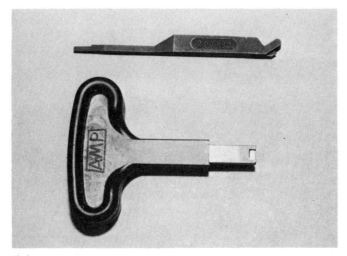

(a)

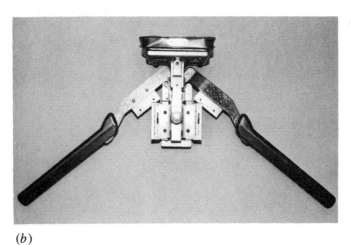

(b)

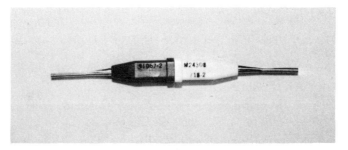

(c)

Fig. A-5 Connector tools. (a) AMP 50-pin connector tools—extraction tool (top, P/N 230089-1), insertion tool (bottom, P/N 229384-1); (b) AMP MI-1 "butterfly" tool (P/N 229378-1); (c) AMP EIA connector pin insertion/extraction tool (P/N 91067-2).

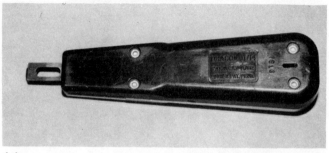

(a)

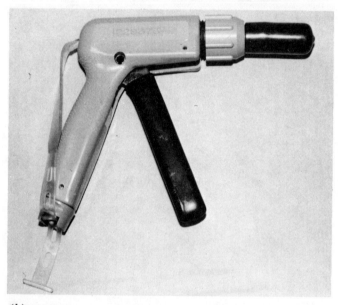

(b)

Fig. A-6 Block tools. (a) 66B3 punch block tool (type 714); (b) AMP block tool (P/N 229373-4).

b. Block Tools (Figure A-6):
- The type 714 tool shown in Figure A-6(a) is used to punch wire on the 66B3 blocks. Try North Supply for this tool.
- The AMP tool shown in Figure A-6(b) is used to terminate stranded wire on AMP punch blocks.

c. Crimp Tools (Figure A-7):
AMP and other manufacturers make several varieties of tools to crimp pins and sockets onto wire. Tools can be purchased which are sized for partic-

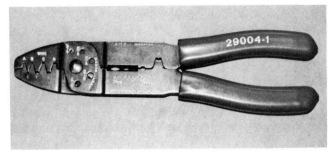

(a)

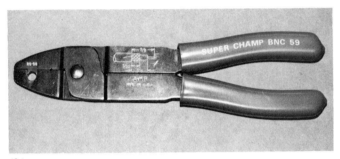

(b)

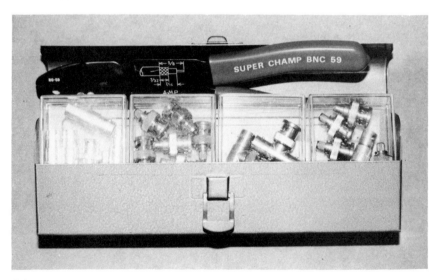

(c)

Fig. A-7 Crimp tools. (a) AMP general-purpose crimp tool (P/N 29004-1); (b) AMP coax crimp tool (P/N 601834-2); (c) AMP coax connector installation kit (P/N 1-601834-3).

ular pins and sockets and can become quite costly. The tool shown in Figure A-7(a) is a compromise. It is a general-purpose tool useful for several sizes of pins and sockets, including all those mentioned in this text. The coax crimp tool shown in Figure A-7(b) is used to install BNC connectors onto coaxial cables. The particular tool shown is the proper one for RG-62 cable used for 327X coaxial networking. Another useful item is the coax connector kit shown in Figure A-7(c). The tool and an assortment of connectors are packaged together. They provide an especially good way to get started if you are unfamiliar with terminating coaxial cable.

  d. Miscellaneous Tools:
- Wire Pulling Sock: This is more formally known as a wire pull grip. Try your local electrical distributor. Kellens is a manufacturer.

VENDOR INFORMATION

Allied Electronics   Try (817) 336-5401 in Fort Worth if you can't find a local Allied outlet.

Alpha Wire   Sells through local electronics distributors.

AMP Special Industries   Sells through local AMP outlet. Call AMP at (717) 564-0100 (Harrisburg) if you need help finding a local rep.

Amphenol   Sells through local electronics distributors.

AT&T
Information Distribution Center
Technical References
60 Kingsbridge Road
Piscataway, NJ 08854

Avery labels   Available from local stationery stores.

Belden   Sells through local electronics distributors.

Bishop Graphics, Inc.
5388 Sterling Center Drive
Westlake Village, CA 91359
(213) 991-2600

Brady Co.
2221 West Camden Road
P.O. Box 2131
Milwaukee, WI 53201
(414) 332-8100

APPENDIX **195**

California Chassis
10636 Midway Ave.
Cerritos, CA 90701
(213) 924-8802

Cinch (TRW)   Sells through local electronics distributors.

Comm-Fab Inc.
16222 E. Arrow Highway
Irwindale, CA 91706
(213) 962-3291

Computer Cable & Products
150 Schmitt Blvd.
Farmingdale, NY 11735
(516) 293-1610

Dakota Engineering Inc.
4315-4317 Sepulveda Blvd.
Culver City, CA 90230
(213) 870-7983

Diamond
500 North Ave.
Garwood, NJ 07027
(201) 789-1400

Electronic Industries Association
Engineering Dept.—Standard Sales
2001 Eye St. N.W.
Washington, DC 20006

E-Z Letter Quik Stik   Available from local stationery stores.

Inmac
2465 Augustine Drive
Santa Clara, CA 95051
(408) 727-1970

McMaster-Carr
P.O. Box 54960
Los Angeles, CA 90054

The Mica Corporation
4031 Elenda St.
Culver City, CA 90230

North Supply Co.
10951 Lakeview Avenue
Lenexa, KS 66219
(800) 255-6889

Panduit Corp.
17301 Ridgeland Ave.
Tinley Park, IL 60477
(312) 532-1800

The Siemon Company
91 Depot Street
Watertown, CT 06795
(203) 274-2523

TW Comcorp
122 Cutter Mill Road
Great Neck, NY 11021
(516) 482-8100

Vari-Tronics Co.
2745 E. Huntington Drive
Duarte, CA 91010
(213) 359-8321

Wiremold Available from local electrical distributors.

Wright Line Inc.
160 Gold Star Blvd.
Worchester, MA 01606

INDEX

66B3 blocks, 189
232/449 adapter, 190
327X cable, 186, 194
327X port documentation, 148

Adapter assemblies, 102, 104, 180
   carrier-controlled CTS, 105
   FDM, 105
   RS-232/449, 190
   sync noise reducer, 106
Agreements, 128
AMP blocks, 193
Archiving, 130, 138
Backup maintenance, 166
Baseband coax, 16
Blocks (*see* Terminal blocks)
Blueprints, 151
Boots, cable connector, 52, 189
Breaker panels, 74
   labeling, 74, 76
Bridge clip blocks, 83
Broadband coax, 16
Bushings for conduit, 30, 31, 38, 190

Cable:
   coaxial, 15, 49, 186
   dressing, 80, 83
   Ethernet, 15
   extender, 100
   stranded, 186, 187
   terminal drop, 58, 186
   twisted pair, 14, 15, 48, 49, 52, 56, 111, 112, 186
   (*See also* Feeder cable)
Cable access doors, 41, 190

Cable assemblies, 97, 99 181, 183
   four-wire extender, 100
   long-haul to short-haul, 103
   MUX to 103(113), 102
   RTS async, 101
   standard async, 100
   sync, 101
   ten-wire extender, 100
   (*See also* Terminal drops)
Cable bar, 86
Cable clamps, 182
Cable connector boots, 52, 189
Cable distribution, 77
Cable dressing, 80, 83
Cable grease, 189
Cable guard, floor, 61, 189
Cable hangers, 42, 56, 77, 78, 181
Cable labeling, 149
Cable raceway, 33
Cable tray, 36, 37, 51, 190
   above ceiling, 36, 39
   below ceiling, 36
   crossovers, 36, 37
   open, 45
Cable tying, 49
Change notice, 134
Checklists:
   for documentation, 180
   for facility, 177
   for networking, 174
Circuit documentation, 137, 140–147
Circuit labeling, 65
Coaxial cable, 15, 49, 186
Coaxial cable patch panel, 24, 95, 185
Coaxial circuit documentation, 139, 146
Coaxial connector insulation, 50, 51, 53, 57

Coaxial feeder:
  bundle sizing, 49
  cable, 23, 49, 53, 97
  cable tying, 49
  standard lengths, 50
Coaxial networking:
  concepts, 23, 24
  physical boundaries, 25
Color coding:
  of EIA signals, 114
  mnemonics for, 113
  resistor, 113
  SEND/RECV, 121
  of telco cable, 111, 112
Comm area design, 72
Conduit:
  bushings for, 30, 31, 38, 190
  cable loading, 44, 57
  for data cable, 36
  flexible, 74, 76
  hallway termination, 31, 38
  installation, 43
  length labeling, 44, 67
  plastic, 57
Connectors:
  25-pin, 188
  50-pin, 52, 187
  coax, 53, 188
  power, 75, 188
  RS-232, 188
  twist-lock, 74, 75, 188
Consultants, 170
Conventions:
  block panel wiring, 114, 117
  reasons for, 110
  SEND/RECV, 120, 122
Crossovers, 36, 37
Curve radius, conduit, 43

D-rings, 183
Data circuit performance, 123
Data comm form, 131
Data comm records, 136
DB-25P/S connectors, 188
Decision making, 169
Diagnostic equipment, 93
Dial circuit documentation, 140
Documentation:
  checklist for, 180
  circuit, 137, 140–147

Documentation (*Cont.*):
  coaxial circuit, 139, 146
  feeder, 149
  maintenance, 164
  network, 137, 150
  port, 145, 147, 148
  power, 152, 154, 155
  twisted pair, 138, 147
Doors, cable access, 41, 190
Dressing cable, 80, 83
Drip loop (*see* Service loops)

EIA signals:
  cabling, 77
  color code, 114
  connectors, 188
  fake out, 94
  mnemonics, 115
EIA specifications, 129
  RS-232 mnemonics, 115
  RS-422, 18
  RS-423, 18
  RS-449 mnemonics, 115
Equipment housing, 82
Ethernet cable, 16
Extender cable:
  four-wire, 100
  ten-wire, 100

Facility checklist, 177
FDM adapter assembly, 105
Feeder cable:
  coaxial, 23, 49, 53, 97
  gender determination, 53
  ID numbers, 64
  installation, 50, 55
  labeling, 62
  linkage, 52
  twisted pair, 48, 52
Feeder documentation, 149
Fire detectors, 72
Flex conduit, 74, 76
Floor outlets, 34, 35
Floor tiles, cutting standards, 122
Forms:
  data comm, 131
  TP change request, 132–134
  work-in-progress, 134, 135

G-10 epoxy board, 185
Glossary, 2
Grease, cable, 189

Halon zones, 72
Hangars, 77, 181

In-house circuit documentation, 144
Insulation, 189
  of coax connector, 50, 51, 53, 57
International Organization for
    Standardization (ISO), 159
Ionization detectors, 72

KV track, 86, 182

Label holders, 182
Labeling:
  of breaker panels, 74, 76
  of cable, 149
  of circuits, 65
  of conduit length, 44, 67
  of equipment, 151
  of feeders, 62
  of patch panels, 97
  of punch block, 115
  of rack-mounted equipment, 86
  of shelf and rack, 86, 150
  of terminal block tabs, 120
  of terminal blocks, 65
  of terminal drops, 66
  of termination panels, 67
  of unused circuits, 66
Labeling materials, 67, 186
Lettering, 186
Library materials, 127
Long-haul circuit documentation, 142

Maintenance:
  backup, 166
  documentation, 164
  in-house, 166
  preventive, 164
  program, 162
  spares, 167
  vendor, 164, 165

Mnemonics, 123
  color coding, 113
  EIA signal, 115
  RS-232, 115
  RS-449, 115
Modem removal, 90

Naming conventions, punch block, 115, 118
Network:
  documentation of, 137, 150
  topologies of, 17
Network access zones, 20
Networking:
  coaxial cable, 23–25
  checklist for, 174
  no-modem, 90
  twisted pair, 19–20
No-modem, synchronous, 95
No-modem networking, 90
No-modem panels, 91
No-modem signal fake out, 95
No-modem terminal drop, 98
Noise considerations, 58
Noise reduction:
  synchronous adapter, 106
  via coaxial cable, 15
  via twisted pair, 14
Noise suppression, 103

Office:
  design for data, 30, 31
  terminal drop for, 58
  terminal drop length for, 59
  wiring for, 58
Outlet plate, 59, 189
Outlets, floor, 34, 35
Overvoltage suppression, 93, 106, 107, 189

Panels:
  breaker, 74, 76
  coax patching, 95, 96
  local data circuit, 93, 185
  no-modem, 91
  space utilization, 123
  terminal block, 182
  wiring, 21, 22, 82
Patch panels, coax, 24, 95, 185
  labeling, 97

Patching wire, 186
Performance of data circuits, 123
Personnel, 158
Policies, 171
  and procedures, 127
Port documentation, 145, 147, 148
Power connectors, 75, 188
Power documentation, 152, 154, 155
Power grid, 74, 154
Power strips, 86, 182
Preventive maintenance, 164
Pull sock, 194
Pullwire, 45, 57, 189
Punch block labeling, 115
Punch block tab numbering, 119
Punch block (*see* Terminal blocks)

Raceway, 33, 35, 39, 60, 61, 189
Rack-mounted equipment, 86
Raised floor, 72
Raised-floor terminal rooms, 32
Records, data comm, 136
Resistor color code, 113
Responsibility of network staff, 168
RG-62 cable, 186, 194
RS-232 mnemonics, 115
RS-422, 18
RS-423, 18
RS-449 mnemonics, 115

Safety considerations, 34
Self-extinguishing blocks, 183
Service loops, 42, 50, 55–57, 62
Shelf labels, 86
Shelving, 84, 181
Shielded twisted pair cable, 15, 107, 108
Short-haul circuit documentation, 141
Signal coupling, 48
66B3 blocks, 189
Spares, 167
Spares rotation, 167
Splicing coax, 51
Spring for cables, 85, 86, 182
Staff, 158
  responsibility of, 168
Strain relief, 42, 58
  for vertical cable, 40
Stranded cable, 186, 187
Stranded twisted pair, 58

Stranded wire, 74
Sync noise reducer, 106

Technical references, 128
Telco color coding, 111, 112
Telephone dial circuits, 72
Terminal block labeling, 65
Terminal block panels, 182
Terminal blocks, 53, 86, 91, 183, 189
  for stranded wire, 22, 54, 58
  tab numbering, 120
Terminal drops:
  cable, 58, 186
  labeling, 66
  office, 39
  RTS async, 101
  standard async, 100
  sync, 101
  (*See also* Cable assemblies)
Terminal-end design concepts, 29
Terminal lab drop length, 60
Terminal labs, 33, 34, 60
  design of, 32
  safety considerations for, 34
Terminal strip, 93
Termination board, 42
  interbuilding, 42, 44
  placement considerations, 43
Termination panel labeling, 67
327X cable, 186, 194
327X port documentation, 148
Tools, 190–194
Topologies of networks, 17
TP change request form, 132–134
Track, KV, 86, 182
Transient suppression, 93, 106
Transient suppressors, 189
Trays (*see* Cable tray)
Troubleshooting, 163
Twist-lock connectors, 74, 75, 188
Twisted pair cable, 14, 15, 48, 52, 186
  color coding of, 111, 112
  selecting gauge, 49, 56
  shielded, 15, 107, 108
Twisted pair networking:
  access zones, 20
  concepts of, 19
  physical boundaries, 19
Twisted pair panel assembly, 183
Twisted pair port documentation, 138, 147

Twisted pair termination, 80
232/449 adapter, 190

Varistor, 93, 106, 107
Vendor maintenance, 164
  reasons for, 165
Vertical cable installation, 40–42
Voltage-dependent resistor (VDR), 93, 106, 107

Wire:
  patching, 186
  stranded, 74
  (See also Cable)
Wire loops, 183
Wire pull grip, 194
Wiring blocks *(See* Terminal blocks)
Wiring panels, 21, 22, 82
Work-in-progress form, 134, 135

ABOUT THE AUTHOR

PAUL S. KREAGER, P.E., is network manager for the Washington State University Computing Center and president of Computer Energy, Inc., a firm providing services and products to the computer/data communications industry. He has more than 20 years' experience in the field of computers and data processing, including operations, hardware maintenance, hardware and software design and engineering, and management. The author of numerous papers in professional journals, Mr. Kreager is also a frequent speaker at seminars and conferences. He received an M.S. in computer science from Washington State University.